2nd International Symposium on Green Technology for Value Chains (GreenVC 2017)

IOP Conference Series: Earth and Environmental Science
Volume 160

Jakarta, Indonesia
23 - 24 October 2017

Editor:

Myrtha Karina

ISBN: 978-1-5108-6478-8
ISSN: 1755-1307

Printed from e-media with permission by:

Curran Associates, Inc.
57 Morehouse Lane
Red Hook, NY 12571

Some format issues inherent in the e-media version may also appear in this print version.

Printed by Curran Associates, Inc. (2018)

For permission requests, please contact the Institute of Physics at the address below.

Institute of Physics
Dirac House, Temple Back
Bristol BS1 6BE UK

Phone: 44 1 17 929 7481
Fax: 44 1 17 920 0979

techtracking@iop.org

Additional copies of this publication are available from:

Curran Associates, Inc.
57 Morehouse Lane
Red Hook, NY 12571 USA
Phone: 845-758-0400
Fax: 845-758-2633
Email: curran@proceedings.com
Web: www.proceedings.com

Table of contents

Volume 160

2nd International Symposium on Green Technology for Value Chains 2017 (GreenVC 2017)

23–24 October 2017, Jakarta, Indonesia

Accepted papers received: 31 May 2018
Published online: 18 June 2018

Preface

Papers

Green Materials and Process

Waste Treatment and Management

Green Energy

Socio-Technological Assessment on Green Production

Life Cycle and Environmental Monitoring

IOP Conf. Series: Earth and Environmental Science **160** (2018) 011001 doi:10.1088/1755-1315/160/1/011001

Preface

Following the successful of the first event, the second International Symposium on Green Technology for Value Chains (GreenVC) 2017 was held on 23-24 October 2017 in Balai Kartini, Jakarta, Indonesia.

This symposium brought together leading academic scientists, researchers, university students and scholars as well as industries, to promote scientific cooperation among participants.

The symposium was organized by the Research Unit for Clean Technology, Indonesian Institute of Sciences (LIPI) and mainly supported by Deputy for Engineering and Sciences, LIPI

The proceedings of GreenVC 2017 presents the original and unpublished results in all areas of interests. Twenty-four papers have been selected based on the originality and interests to the audiences. Topics on (1) Green Materials and Process, (2), Waste Treatment and Management (3), Green Energy, (4) Socio-technological Assessment on Green Production, and (5) Life Cycle and Environmental Monitoring were presented orally or in poster presentation.

We would like to express our sincere gratitude to our Advisory Board, Scientific and Editorial Board for their positive contribution to keep the quality of the articles presented in this proceeding. We also would like to appreciate all participants attending this symposium as well as to Organizing Committee who made the symposium successful.

The Third International Symposium on Green Technology for Value Chains will take place in BSD City, Indonesia, on 1-2 November 2018.

May 30, 2018

Editor in Chief
Prof. Myrtha Karina

IOP Conf. Series: Earth and Environmental Science **160** (2018) 011001 doi:10.1088/1755-1315/160/1/011001

Organizing Committee

Advisory Board

- Dr. Laksana Tri Handoko (Indonesian Institute of Sciences, Indonesia)
- Prof. Hiroshi Uyama (Osaka University, Japan)
- Prof. Akkharawit Kanjana-Opas (Prince of Songkla University, Thailand)

Scientific Committee / Editorial Board

- Prof. Myrtha Karina Sancoyorini (Indonesian Institute of Sciences, Indonesia)
- Dr. Rike Yudianti (Indonesian Institute of Sciences, Indonesia)
- Dr. Neni Sintawardani (Indonesian Institute of Sciences, Indonesia)
- Prof. Tarzan Sembiring (Indonesian Institute of Sciences, Indonesia)
- Dr. Hari Rom Hariyadi (Indonesian Institute of Sciences, Indonesia)
- Dr. Muhammad Nasir (Indonesian Institute of Sciences, Indonesia)
- Dr. Sri Pudjiraharti (Indonesian Institute of Sciences, Indonesia)
- Dr. Tri Muji Ermayanti (Indonesian Institute of Sciences, Indonesia)
- Dr. Nanik Indayaningsih (Indonesian Institute of Sciences, Indonesia)
- Dr. Euis Hermiati (Indonesian Institute of Sciences, Indonesia)
- Dr. Maman Rahmansyah (Indonesian Institute of Sciences, Indonesia)
- Dr. Haznan Abimanyu (Indonesian Institute of Sciences, Indonesia)
- Prof. Silvester Tursiloadi (Indonesian Institute of Sciences, Indonesia)
- Dr. Oman Zuas (Indonesian Institute of Sciences, Indonesia)
- Dr. Tjandra Mozef (Indonesian Institute of Sciences, Indonesia)
- Dr. Ajeng Arum Sari (Indonesian Institute of Sciences, Indonesia)
- Prof. Dr. Yanni Sudiayani (Indonesian Institute of Sciences, Indonesia)
- Dr. Ir. Heddy Julistiono (Indonesian Institute of Sciences, Indonesia)
- Dr. Anastasia Fitria Devi (Indonesian Institute of Sciences, Indonesia)
- Prof. Naoyuki Funamizu (Hokkaido University, Japan)
- Prof. Dr. M. Djaeni (Diponegoro University, Indonesia)
- Dr. Lina Abu Ghunmi (University of Jordan, Jordan)
- Dr. Shahrul Ismail (Universiti Malaysia Terengganu, Malaysia)
- Dr. Shameen Jinadasa (University of Peradeniya, Sri Lanka)
- Dr. Mariam Sou Dakoure (International Institute for Water and Environmental, Burkina Faso)
- Dr. Ken Ushijima (Hokkaido Research Organization, Japan)
- Tjahjo Tamtomo, PhD (University of Indonesia, Indonesia)
- Dr. Ahmad Komarulzaman (Padjadjaran University, Indonesia)

Organizer

- Dr. Widyarani (Chair)
- Dr. Indah Primadona
- Dr. Desak G. Sri Andayani
- Elsy Rahimi Chaldun, M.T.
- Indriyati, M.Eng
- Dr. A. Hanif Dawam Abdullah
- Dr. Fitri Dara
- Panji Cahya Mawarda, MSc.
- Robby Wahyu Sophian

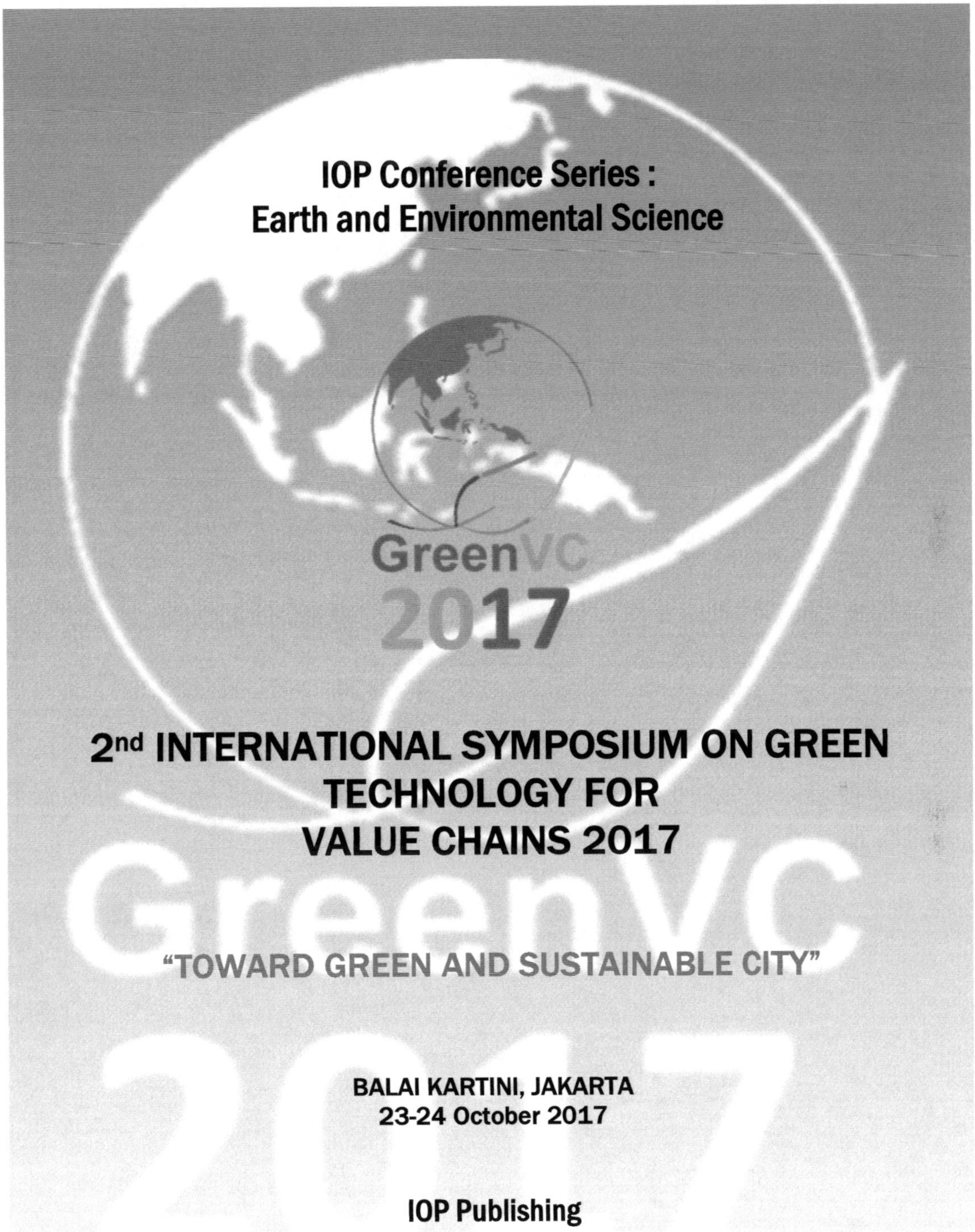
IOP Conference Series :
Earth and Environmental Science
GreenVC
2017
2nd INTERNATIONAL SYMPOSIUM ON GREEN
TECHNOLOGY FOR
VALUE CHAINS 2017
GreenVC
"TOWARD GREEN AND SUSTAINABLE CITY"
2017
BALAI KARTINI, JAKARTA
23-24 October 2017
IOP Publishing

IOP Conf. Series: Earth and Environmental Science **160** (2018) 011002 doi:10.1088/1755-1315/160/1/011002

Peer review statement

All papers published in this volume of *IOP Conference Series: Earth and Environmental Science* have been peer reviewed through processes administered by the proceedings Editors. Reviews were conducted by expert referees to the professional and scientific standards expected of a proceedings journal published by IOP Publishing.

Characteristic of nanoparticle-chitosan system: solution and thin film study

M Nasir* **and R I Sugatri**

Research Unit for Clean Technology, Indonesian Institute of Sciences (LIPI)
Komplek LIPI Bandung, Jalan Sangkuriang, Gedung 50, Bandung 40135, Indonesia

*Email: mnasir71@yahoo.com

Abstract. Nanoparticle (NP)-chitosan system is very important composite nanomaterial related to their unique properties and applications. In this work, chitosan was composited by two types of nanoparticle i.e. silver nanoparticles (Ag (NP)) and activated nanocarbon (ANC), respectively. The presence of nanoparticle in chitosan system gave effect to viscosity of solution and hydrophilicity of thin film. In Ag (NP)-chitosan system, the addition of 10, 15 and 25% (v/v) Ag (NP) suspension increased their viscosity. Different result was obtained for ANC-chitosan system,where addition of 0.1, 0.5 and 1.5% (w/v) of ANC deacreased viscosity of solution.This property affected distribution of the nanoparticles in chitosan thin film and its hydrophilicity. The chitosan thin film nano-compositehad higher hydrophilicity than pure chitosan. Infrared spectrum analysis showed that nanoparticles which were composited in chitosan system gave shifting in O-H stretching vibration, C=O stretching vibration (amide I) and N-H bending and C-N stretching (amide II).

1. Introduction

Chitosan (β (1→4) links 2-amino-2-deoxy-D-glucopyranose) is the most important derivate of chitin which is soluble in acidic aqueous media and is used in many applications such as food, cosmetic, biomedical and pharmaceutical applications [1]. Chitosan has antibacterial activities and it has been proven excellent building blocks to produce films with controlled molecular architecture, allowing for synergy between distinct materials [2]. Linear structure of this polysaccharide renders their film tough, flexible and transparent [3]. Chitosan thin film is used as one of the food packaging material in biopolymer [4] and antibacterial [5]. To improve pure chitosan film properties, it can be modified by adding distinct materials such as nanoparticle or other polymers.

The composite of chitosan with other substance such as CuO, Cu nanoparticle, Ag nanoparticle, nanocarbon, carbon nanotubes, graphene oxide, montmorillonite, ZnO and TiO_2[5–16] will give effect to its function and characteristics. The addition of metal particles such as silver or copper into polymer produced novel antimicrobial materials [17]. It was reported by Tamayo *et al* [18] that copper-polymer nanocomposite showed excellent antimicrobial properties. The addition of CuO in chitosan system enhanced tensile strength and stiffness [6]. Furthermore, CuO-chitosan thin film can be used as a photocatalyst to degrade organic contaminants [7] and enhanced the antibacterial activities [10]. Raghavendra *et al* [5] combined two different materials such as chitosan and Ag nanoparticles. The antibacterial properties of Ag-chitosan nanocomposite films more effective inhibited gram-positive and gram-negative microorganisms than pure chitosan. Montmorillonite-CuO composite was successfully used to modify of the chitosan film. The composite film showed intense antibacterial activity and improved mechanical properties more than 50% compared with pure chitosan [19]. In nanocarbon system, modified carbon nanoparticle into chitosan film can be used for application in sensing, electrocatalysis, or adsorption [13]. In these cases, chitosan was used in many applications due to abundant chelating groups including primary and secondary hydroxyl groups as well as highly reactive amino group.

In this work, we combined nanoparticle i.e. silver nanoparticle (Ag (NP)) and activated nanocarbon (ANC) into chitosan solution, respectively, to produce nanoparticle-chitosan thin film. The presence of other materials in chitosan film gave effect to their characteristics such as contact angle, the porous of the film and roughness [4,5,20].

The characteristic of nanoparticle-chitosan thin films composite was observed including the viscosity of solution composite, fluorescence microscope of thin film to observe distribution of nanoparticle in chitosan system, IR spectrum, the surface morphology by Scanning Electron Microscopy (SEM) and water contact angle.

2. Experimental

2.1. Materials

Chitosan medical grade (94.50% deacetylated) was obtained by PT. Biotech Surindo. Tri-sodium citrate dehydrate and silver nitrate were purchased from Merck. Carbon granule, potassium hydroxide and acetic acid glacial were purchased from PT. Brataco.

2.2. Synthesis of silver nanoparticle (Ag (NP))

An amount of 1mM silver nitrate and 1% (w/v) tri-sodium citrate with ratio 10:1 (v/v) were dissolved in distilled water, respectively then mixed. The mixture was heated in a modified microwave oven (Panasonic, 900W) which was connected with refluxing system in low condition for 5 min.

2.3. Activation of nanocarbon

Carbon granule was grinded, sieved and then milled by a High Energy Milling E3D (HEM-E3D) for 25 h to obtain nanocarbon. Nanocarbon was activated by using 20% (w/v) potassium hydroxide which had been dissolved in distilled water with ratio nanocarbon:KOH solution 1:4 (w/v). The mixture was stirred for 24 h at room temperature, neutralized with distilled water, then oven-dried.

2.4. Synthesis of nanoparticle-chitosan system

2.4.1. Ag (NP)-chitosan solution. Ag (NP)-chitosan solution was made from Ag (NP) suspension in chitosan system of5, 15, and 255% (v/v). The 2.5 g of chitosan (2.5% w/v) was dissolved in 75, 85 and 95 mL acetic acid (2% v/v), respectively. The mixture was stirred for 12 h at room temperature. The Ag (NP) suspension was added into chitosan solution 5 mL, 15 mL and 25 mL, respectively, then stirred for 1 h at room temperature.

2.4.2. ANC-chitosan solution. ANC-chitosan solution was made from ANC in chitosan system of 0.1, 0.5 and 1.5% (w/v). The 2.5 g of chitosan (2.5% w/v) was dissolved in 100 mL acetic acid (2% v/v) followed by stirring for 12 h at room temperature. The ANC of 0.1, 0.5 and 1.5 g were added into 100 mL of chitosan solution, respectively, stirred for 1 h then sonicated for 1 h.

2.4.3. Preparation of thin film. All bubbles in the nanoparticle-chitosan were eliminated before casting onto mold which was made up of flat glass and coated by aluminium foil. The solutions were poured into mold and allowed to dry at 50-60°C for 12 h.

2.5. Characterization

The viscosity of nanoparticle-chitosan solution was determined by Brookfiled Viscometer DV-I Prime. The thin film composite was characterized by fluorescence microscope Meiji Techno America Japan to observe distribution of nanoparticle in chitosan system. Fourier Transform Infrared Spectroscopy (FTIR) Shimadzu IR Prestige 21 was used to analyse the interaction of nanoparticle and chitosan. Surface morphology was analyzed by Scanning Electron Microscopy (SEM) SEMSU3500. Calculating water contact angle used the equation: equation: $\frac{\theta}{2} = \tan^{-1}\left(\frac{h}{d}\right)$ with Θ = contact angle, d = diameter, h = height. Pure chitosan thin film was characterized for comparison.

3. Results and discussion

Chitosan membrane is the ability of beta-(1,4)-linked D glucosamine with certain degree deacetylation to re-form supramolecular structures via hydrogen bonding after evaporation of the solvent [20]. Chitosan thin films composite were developed by blending nanoparticle and chitosan with varied concentration.

3.1. Viscosity analysis

Silver nanoparticles (Ag (NP)) were synthesized by microwave method. The method can produce Ag which has smaller particles and high stability with faster reaction [21–26]. In the nanoparticle-chitosan system, nanoparticles were added into chitosan solution to disperse nanoparticles in chitosan matrix. The addition of difference of nanoparticles into chitosan solution gave a different effect on viscosity.

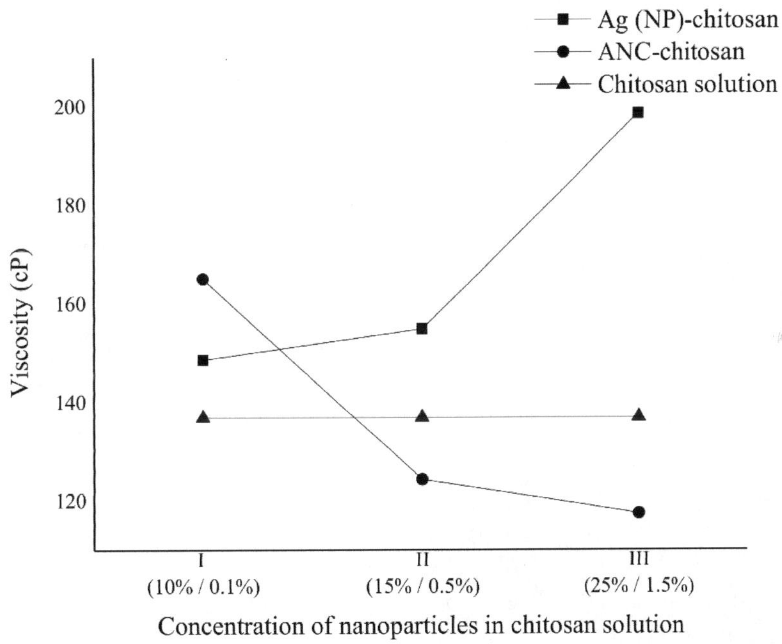

Figure 1. Correlation between concentration of nanoparticle-chitosan solution and viscosity.

In Figure 1, it showed that the addition of Ag (NP) suspension in10%, 15%, and 25%-chitosan solution enhanced viscosity value to 148.60, 154.85, and 198.55 cP for Ag (NP), respectively compared with pure chitosan solution of 136.90 cP. This was caused by the well dissolution of mixture consisted of acidic chitosan solution (pH < 6.5) and acidic Ag (NP) suspension (pH 5.23 to 6.25) [27]. Alfrey *et al* [28] explained that the polymer in a favorable solvent will exhibit a high intrinsic viscosity. The long chain molecule of the polymer will be surrounded by a solvated hull which tends to prevent polymer-polymer contacts. Uncurled configurations are favored, and the mean value of any property represents a more extended shape than the unbiased mean. Since an extended of uncurled configuration is associated with a high intrinsic viscosity.

IOP Conf. Series: Earth and Environmental Science **160** (2018) 012001 doi:10.1088/1755-1315/160/1/012001

(a) Chitosan

(b) Ag (NP) 10%-chitosan

(c) Ag (NP) 15%-chitosan

(d) Ag (NP) 25%-chitosan

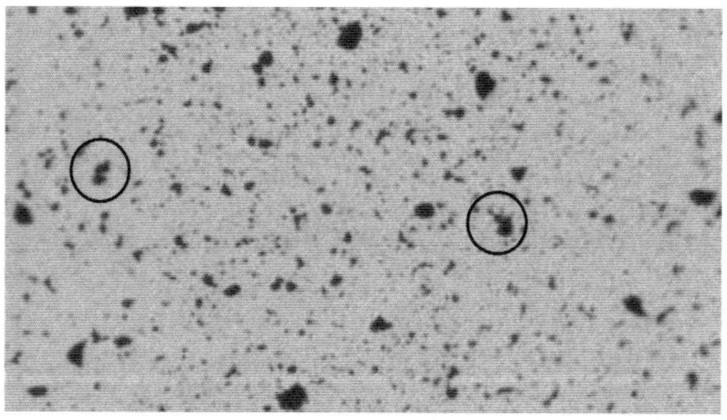

(e) ANC 0.1%-chitosan

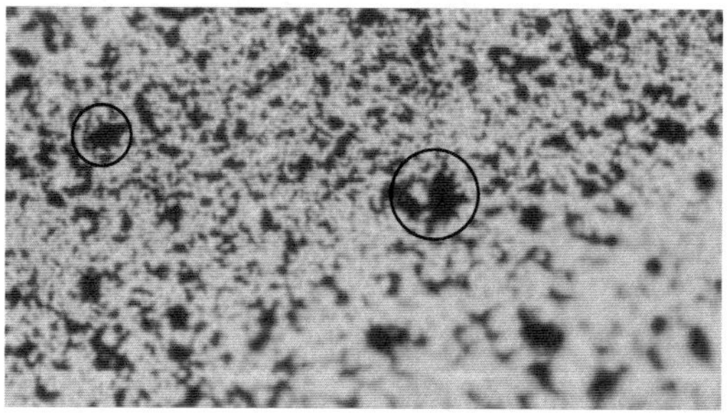

(f) ANC 0.5%-chitosan

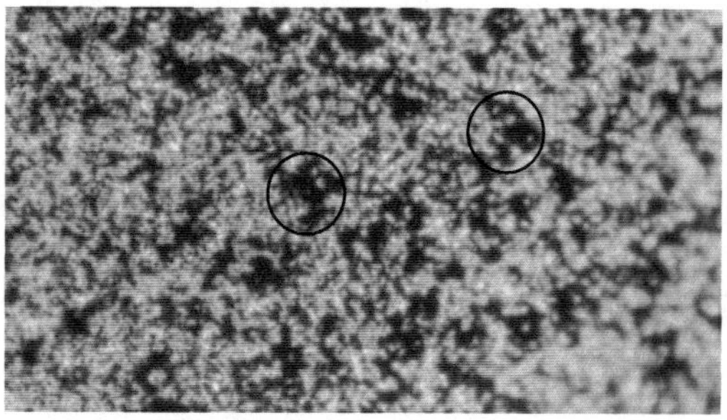

(g) ANC 1.5%-chitosan

Figure 2. Fluorescence microscope image of nanoparticle-chitosan thin film composite (100x).

IOP Conf. Series: Earth and Environmental Science **160** (2018) 012001 doi:10.1088/1755-1315/160/1/012001

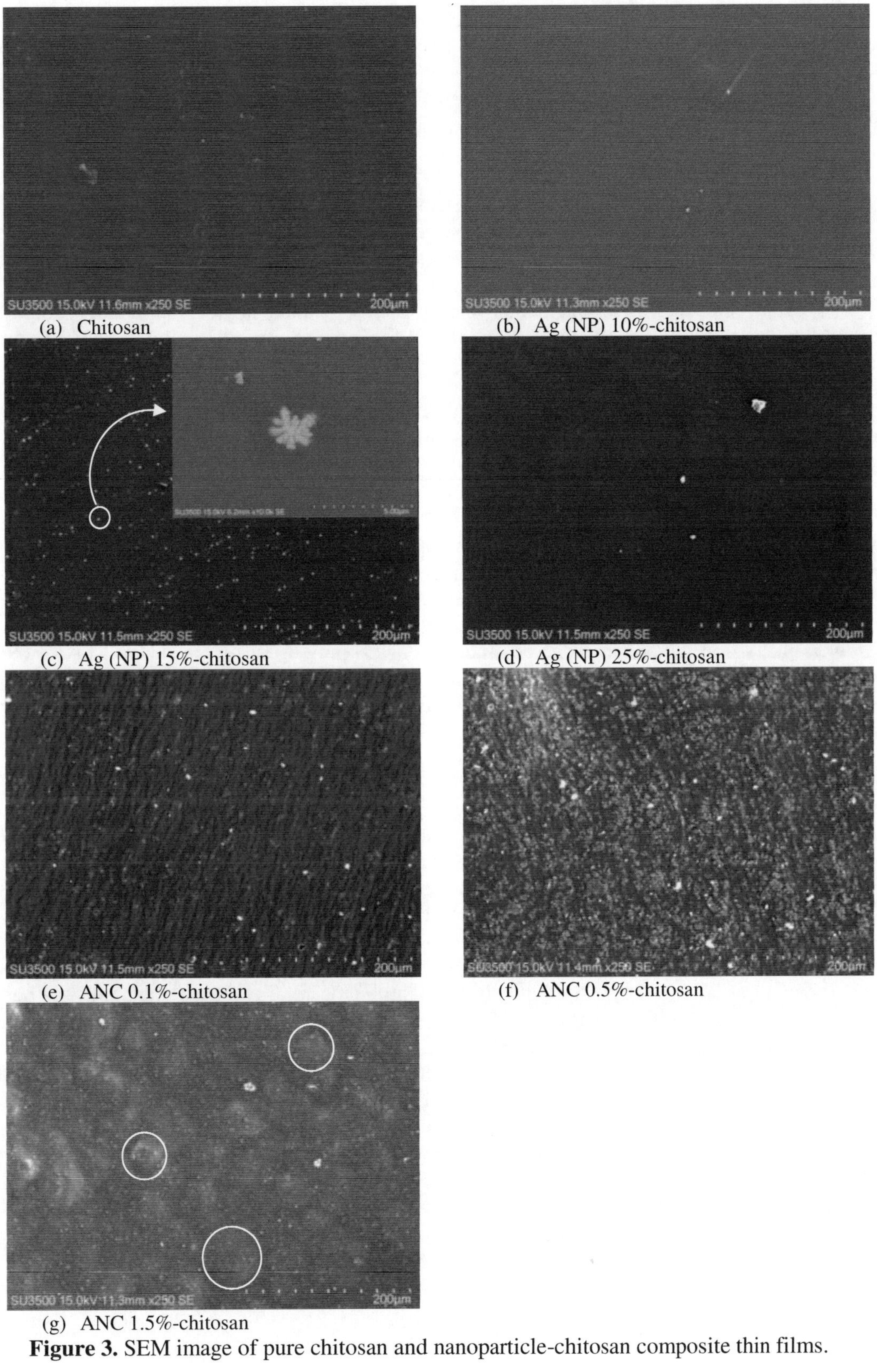

(a) Chitosan

(b) Ag (NP) 10%-chitosan

(c) Ag (NP) 15%-chitosan

(d) Ag (NP) 25%-chitosan

(e) ANC 0.1%-chitosan

(f) ANC 0.5%-chitosan

(g) ANC 1.5%-chitosan

Figure 3. SEM image of pure chitosan and nanoparticle-chitosan composite thin films.

Different case occurred in ANC-chitosan solution of which the viscosity of solution composite depends on the concentration of ANC. Volume fraction, heterogeneous, and homogeneous mixtures affected on viscosity [29]. A particle with sizes with greater volume fractions has low viscosity value and viscosity of the heterogeneous mixture is greater than the homogeneous mixture [30]. When ANC was added into chitosan solution, it disperses in the chitosan solution. Nevertheless, ANC-chitosan solution formed a heterogeneous mixture. It shows in Figure 2 some aggregates of ANC in chitosan solution and it did not distribute well. Boylu *et al* [30] reported that high volume fractions in the system provided more void space which caused lower viscosity and greater fluidity. Carbon with low volume fractions exhibits a rather compacted structure, and thus the fluid that forms a suspension with the carbon cannot easily enter the intergranular void spaces in carbon. Even if the fluid enters those void spaces it cannot move or flow freely within the occupied sites. In ANC-chitosan solution, increasing activity of polymer-polymer contacts with lower fraction volume of ANC resulted in greater resistance. This caused increasing its viscosity of 117.40, 124.30, 165.05 cP for 1.5, 0.5, and 0.1% (w/v) ANC in chitosan solution.

3.2. Nanoparticle-chitosan thin film composite

Fluorescence microscope of chitosan and nanoparticle-chitosan thin films is shown in Figure 2. Nanoparticle dispersed in chitosan matrix.

Different from ANC-chitosan film, increasing concentration of nanoparticle in Ag (NP)-chitosan thin film tends to disperse in chitosan matrix well. The morphology of nanoparticle-chitosan thin film composite is shown in Figure 3. SEM image clearly shows the coarse surface of pure chitosan thin film with the presence deposit of chitosan. On the other hand, the surface of Ag (NP)-chitosan thin film composite seems smooth also with aggregates and deposited on the surface.

Figure 3c shows morphology of Ag(NP)-chitosan thin film composite. The composite morphology has flowers-like shape. In this case, the polymer acts as capping agent on the surfaces of Ag (NP) to form flowers-like [31,32]. It was suggested that the morphology of ANC-chitosan thin film composite depends on the addition of ANC into chitosan matrix. The increase of ANC caused the rough surface and tends to form aggregate.

3.3. Infrared spectrum analysis

The chemical compositions of chitosan and the nanoparticle-chitosan composite thin films were examined by FTIR spectroscopy. The spectrum of pure chitosan thin film at 3381.21 cm^{-1} showed the vibration of O-H, at 2933.73 and 2893.22 $cm^{-1}$for bending of C-H, at 1629.85 cm^{-1} stretching vibration of C=O (amide I), at 1546.91 cm^{-1} as bending vibration of N-H and stretching of C-N (amide II), at 1409.96 cm^{-1} bending CH_2 in CH_2OH, 1153.43 and 1091.71 vibration of C-O-C.

In Ag (NP)-chitosan thin film composite, 3381.21 cm^{-1} peak of O-H stretching vibration of pure chitosan was shifted to 3360.00 cm^{-1} and it became broader. The peak at 1629.85 cm^{-1} due to C=O stretching vibration (amide I) was shifted to 1624.06 cm^{-1} and the 1546.91 cm^{-1} peak of N-H bending and C-N stretching (amide II) was shifted to 1554.63 cm^{-1}. The reaction involved the Ag (NP) was bound by amino and hydroxyl group of chitosan chain [33]. The presence of ANC in ANC-chitosan thin film composite caused the lower intensity if compared with pure chitosan thin film. In addition, the shifting to 3385.07, 1620.21, and 1523.76 cm^{-1} showed O-H stretching, C=O stretching (amide I), and N-H bending and C-N stretching vibration (amide II), respectively.

Nanoparticles existence in chitosan system caused shifting in O-H stretching vibration at 3300-4000 cm^{-1}), C=O stretching vibration (amide I), and N-H bending and C-N stretching (amide II), indicated the interaction between nanoparticle with chitosan [34,35].

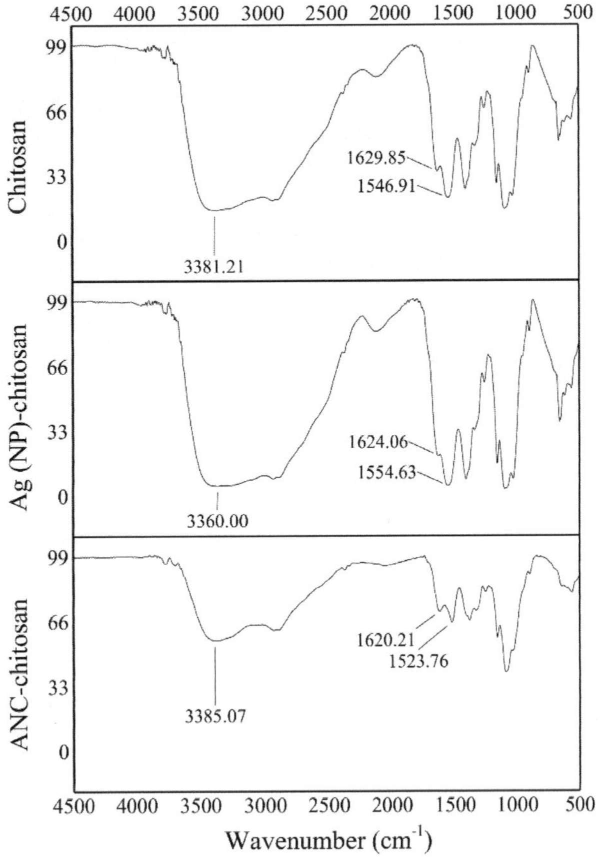

Figure 4. IR spectrum of nanoparticle-chitosan thin film composite.

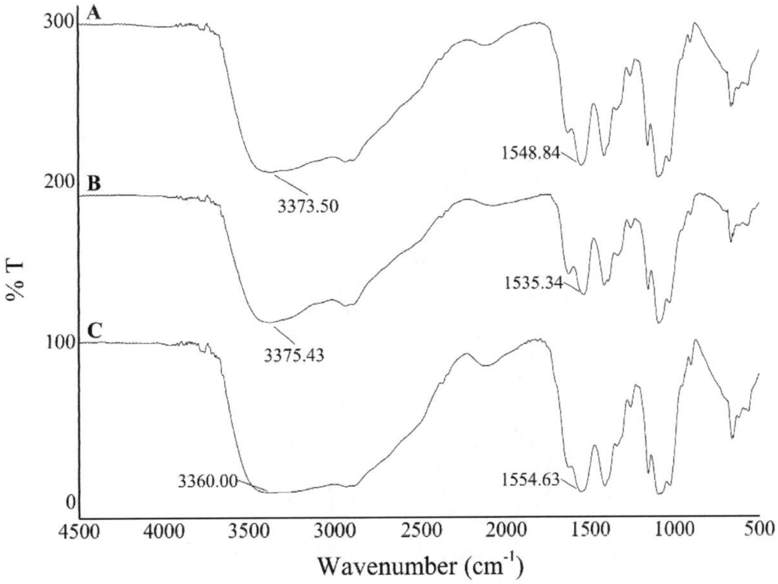

Figure 5. IR Spectrum of Ag (NP)-chitosan with concentration of Ag (NP) (a) 10%, (b) 15%, (c) 25%.

The concentration of nanoparticles which was added into chitosan solution also gave an effect toward its IR spectrum. Figure 5 shows that increasing concentration of Ag (NP) caused shifting to lower wave number. The O-H stretching vibration of pure chitosan was 3381.21 cm^{-1} shifted to 3373.50, 3375.43, and 3360.00 cm^{-1} for addition of Ag (NP) 10%, 15%, and 25%, respectively. Increasing the concentration of nanoparticle also changed the band of O-H stretching vibration. The O-H vibration spectrum becomes wider. The N-H bending and C-N stretching (amide II) vibration band peak of composite also shifted to 1548.84, 1535.34, and 1554.63 cm^{-1} for Ag (NP) 10%, 15%, and 25%, respectively. The vibration band peak of pure chitosan thin film is 1546.91 cm^{-1}. It can be assumed that there was an interaction between the nanoparticle with O-H and N-H in chitosan.

3.4. Contact angle analysis

The hydrophilicity of nanoparticle-chitosan thin film composite was observed by measuring the contact angle of water droplet on the surface of the thin film. Every type of nanoparticles in chitosan matrix gave a different effect on water contact angle.

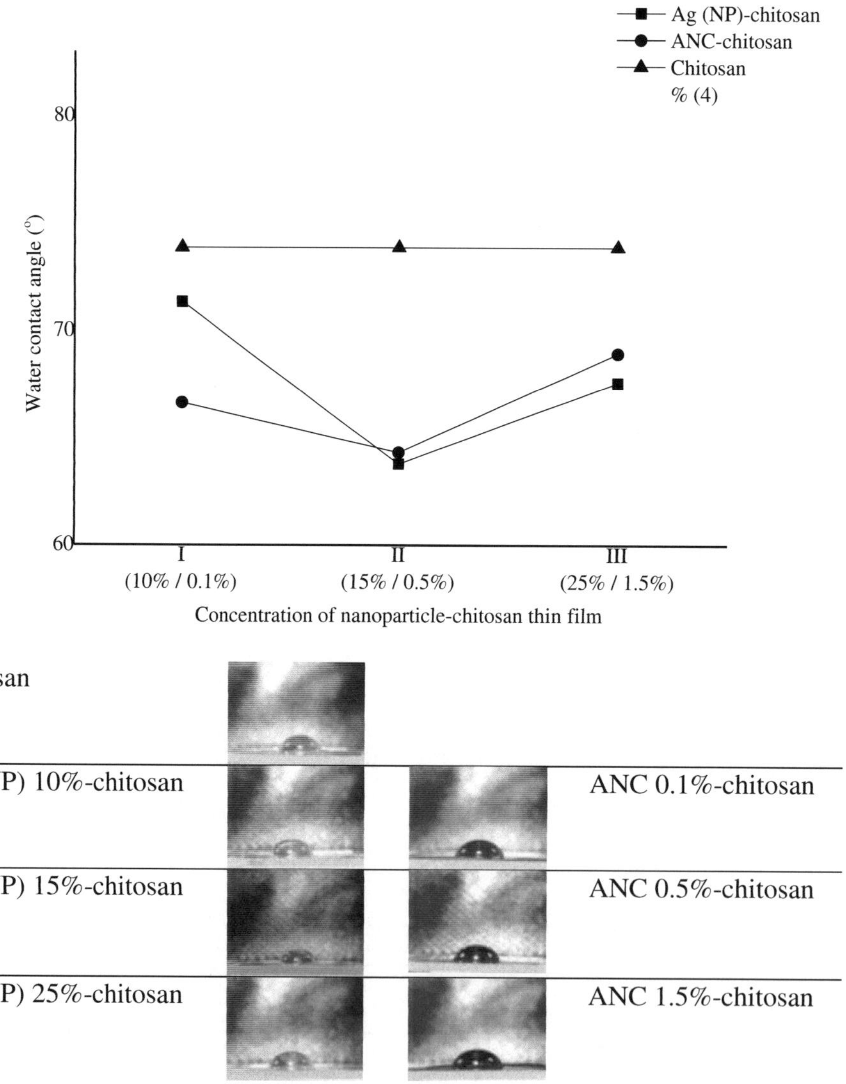

Figure 6. Correlation between concentration of nanoparticle-chitosan thin film and water contact angle.

Pure chitosan thin film is a hydrophilic membrane with a contact angle of water of 73.89°. The addition of nanoparticle in chitosan matrix decreased contact angle of water droplet. In Ag (NP)-chitosan thin film composite, increasing concentration of Ag (NP) in chitosan matrix tended to increase the hydrophilicity. Mohdy *et al* [36] also reported that adding Ag (NP) in polymer increased the hydrophilicity of membrane composite. The water contact angle of Ag (NP) 10, 15, and 25% (v/v) was 71.34, 63.80, and 67.58°, respectively. Nevertheless, inconsistent decrease of contact angle in respective concentration was caused by the imperfect of Ag (NP) distribution in chitosan matrix. The affinity between Ag (NP) and water might be possible to the high hydrophilicity.

On the other hand, as for ANC-chitosan thin film composite, the addition of ANC in chitosan matrix caused the hydrophilicity. The water contact angle of ANC 0.1, 0.5, and 1.5% (w/v) in chitosan matrix was 66.64, 64.33, 68.94°, respectively. Increasing concentration of ANC affected the surface roughness. ANC 1.5%-chitosan shows higher roughness than ANC 0.1%-chitosan thin film composite. The higher surface roughness composite will give higher water contact angle. However, ANC was not distributed well in chitosan matrix (see Figure2), due to high hydrophilicity of ANC 0.5%-chitosan if compared with ANC 0.1%-chitosan. Depends on the concentration and type of nanoparticle, the addition of nanoparticle in chitosan system increased its hydrophilicity.

4. Conclusion

Nanoparticle-chitosan thin film composite was synthesized successfully by mixing nanoparticle and chitosan solution. The concentration and type of nanoparticle affected the viscosity and hydrophilicity. In Ag (NP)-chitosan system, increasing concentration of Ag (NP) in chitosan matrix decreased the viscosity of solution composite and enhanced the hydrophilicity. In ANC-chitosan system composite, increasing concentration of nanoparticles decreased the viscosity of the solution. In addition, the presence of ANC tended to form aggregation on thin film composite and caused roughness as well decreased the hydrophilicity.

Acknowledgments

The authors gratefully acknowledge to Ministry of Research, Technology and Higher Education of the Republic of Indonesia which has funded this research.

References

[1] Rinaudo M 2006 *Prog. Polym. Sci.* **31** 603-32
[2] Pavinnato F J, Caseli L and Oliveira O N 2010 *Biomacromol* **11** 1897-908
[3] Vieira M G A, da Silva M A, dos Santos L O and Beppu M M 2011 *Eur. Polym. J.* **47** 254-63
[4] Sivaselvi K and Ghosh P 2017 *Mater. Today: Proc.* **4** 442-51
[5] Raghavendra G M, Jung J, Kim G and Seo J 2016 *Int. J. Biol. Macromol.* **84** 281-8
[6] Zubair M, Najeeb F, Mohamed M, Saeed A, Ramis M K, Abdul Mujeeb M S and Sarojini B K 2014 *Amer. J. Polym. Sci.* **4** 40-5
[7] Kumar P S, Selvakumar M, Babu S G, Jaganathan S K, Karuthapandian S and Chattopadhyay S 2015 *RSC Adv.* **5** 57493-501
[8] Nayak V, Jyothi M S, Balakrishna R G, Padaki M and Ismail A F 2015 *Chem. Open* **4** 278-87
[9] Al-Naamani L, Dobretsov S and Dutta J 2016 *Innovative Food Sci. Emerging Technol.* **38** 231-7
[10] Widiarti N, Sae J K and Wahyuni S. 2017 *IOP Conf. Ser.: Mater. Sci. Eng.* **172** 012036
[11] Shim I-W, Noh W-T, Kwon J, Cho J Y, Kim K-S and Kang D H 2002 *Bull. Korean Chem. Soc.* **23** 563-6
[12] Cioffi N, Torsi L, Ditaranto N, Tantillo G, Ghibelli L, Sabbatini L, Bleve-Zacheo T, D'Alessio M, Zambonin P G and Traversa E 2005 *Chem. Mater.* **17** 5255-62
[13] Rassaei L, Silanpaa M and Marken F 2008 *Electrochim. Acta* **53**5732-38
[14] Ganesh S, Arockiadoss and Ramaprabhu S 2010 *AIP Conf. Proc.* **1276** 158-62
[15] Qiu J D, Xie H Y and Liang R P 2008 *Microchim. Acta* **162** 57-64
[16] Wang S-F, Shen L, Zhang W-D and Tong Y-J 2005 *Biomacromol.* **6** 3067-72

[17] Palza H, Quijada R and Delgado K 2015 *J. Bioact. Compat. Polym.* **30** 366–80
[18] Tamayo L, Azócar M, Kogan M, Riveros A and Páez M 2016 *Mater. Sci. Eng. C* **69** 1391-409
[19] Nouri A, Yaraki M T, Ghorbanpour M, Agarwal S and Gupta V K 2018 *Int. J. Biol. Macromol.* **109** 1219-31
[20] Clasen C, Wilhelms T and Kulicke W M 2006 *Biomacromol.* **7** 3210-22
[21] Hong Z, Cao Y and Deng J 2002 *Mater. Lett.* **52** 34-8
[22] Volanti D P, Keyson D, Cavalcante L S, Simões, Joya M R, Longo E, Varela J A, Pizani P S and Souza A G 2008 *J. Alloys Compounds* **459** 537-42
[23] El-Trass A, ElShamy H, El-Mehasseb I and El-Kemary M 2012 *Appl. Surface Sci.* **258** 2997-3001
[24] Guo L, Tong F, Liu H, Yang H and Li J 2012 *Mater. Lett.* **71** 32-5
[25] Rejith S G and Krishnan C 2013 *Mater. Lett.* **106** 87-9
[26] Tran T H and Nguyen V T 2014 *Int. Scholar. Res. Notices* **2014** 856592
[27] Mori Y, Tagawa T, Fujita M, Kuno T, Suzuki S, Matsui T and Ishihara M 2011 *J. Nanopart. Res.* **13** 2799-806
[28] Alfrey T, Bartovics A and Mark H. 1942 *J. Am. Chem. Soc.* **64** 1557–60
[29] Elbishari H, Satterthwaite J and Silikas N 2011 *Int. J. Mol. Sci.* **12** 5330-8
[30] Boylu F, Dincer H and Atesok G 2004 *Fuel Proc. Technol.* **85** 241-50
[31] Ali P and Rouhollah S 2011 *Mater. Res. Bull.* **46** 1860-5
[32] Ning Z, Dongsheng L and Deren Y 2014 *Nanoscale Res. Lett.* **9** 302
[33] Mori Y, Ono T, Miyahira Y, Nguyen V Q, Matsui T and Ishihara M 2013 *Nanoscale Res. Lett.* **8** 93
[34] Rahnama A and Gharagozlu M 2012 *Opt. Quant. Electron.* **44** 313-22
[35] Ahmad M B, Lim J J, Shameli K, Ibrahim N A and Tay M Y 2011 *Molecules* **16** 7237-48
[36] Mohdy A, Abou-Okeil A, El-Sabagh S and El-Sawy S M 2015 *Int. Scholar. Sci. Res. Innov.* **9** 1149-54

IOP Conf. Series: Earth and Environmental Science **160** (2018) 012002 doi:10.1088/1755-1315/160/1/012002

The effect of substrate and enzyme concentration on the glucose syrup production from red sorghum starch by enzymatic hydrolysis

A R Permanasari[1*], F Yulistiani[1], R W Purnama[1], T Widjaja[2] and S Gunawan[2]

[1] Politeknik Negeri Bandung, Bandung, Indonesia
[2] Institut Teknologi Sepuluh Nopember, Surabaya, Indonesia

[*]Email: ayu.ratna@polban.ac.id

Abstract. Red sorghum starch contains 19% amylose and 34% amylopectin but low gluten, therefore, it is not suitable for bakery production. On the other hand, red sorghum is potential for glucose syrup production. Glucose syrup can be produced by enzymatic hydrolysis from starch by liquefaction process with α-amylase and saccharification process with glucoamylase. Enzyme concentration plays an important role in the yield of hydrolysis process. This experiment aimed to produced glucose syrup from red sorghum starch by hydrolysis enzymatic method. Liquid α-amylase and glucoamylase were applied for red sorghum starch with particle size 125 μm. Concentrations of α-amylase and glucoamylase were 0.2 and 0.27% (v/v), respectively. Substrate concentration was varied at 10, 20, 30, and 40% (w/v) for each 300 ml substrate solution. Liquefaction process was carried out at 90°C for 60 min whereas saccharification process was carried out at 60°C for 120 min with constant pH 6 for each process. The substrate was then strained as the hydrolysate. The hydrolysate was analyzed for its reducing sugar content by dinitrosalicylic acid (DNS) reagent. The glucose concentration was analyzed by HPLC. The highest reducing sugar concentration was 191.60 g/l, 17.44% glucose observed for 40% substrate and 0.27% enzyme concentration, respectively.

1. Introduction

Nowadays consumption of sugar cane in food and beverage industry is increasing rapidly. To fulfil its high consumption, artificial sweetener is applied recently. Liquid sugar which has less impurities and does not require solubilization is a better alternative for sugar cane crystal substitution. Liquid sugar includes glucose and fructose was produced from some biomass source such as leucalyptus, mahogany sawdust, rice husk [1], water hyacinth [2], starch from rotten potato [3], and cassava (*Manihot esculenta* Crantz) [4].

Sorghum is a staple food crop in some semi-arid tropics region. It has a good potential for commercial exploitation [5]. Starch content of sorghum grain is up to 72% [6] and it is potential for liquid sugar such as glucose or fructose syrup application [6].

Hydrolysis of starch to produce glucose syrup can be obtained with acid hydrolysis e.g. sulfuric acid or hydrochloric acid, enzymatic hydrolysis, or its combination. Enzymatic hydrolysis has fundamental differences with acid hydrolysis. Acid hydrolysis breaks the chain of starch at random parts, whereas enzymatic hydrolysis breaks the chain of starch at certain branching specifically. Enzymatic hydrolysis has some advantages i.e. the process is more specific, the conditions can be controlled, less cost for purification, less ash and by-products, and the color damage can be minimized [7].

Glucose can be made from enzymatic hydrolysis by α-amylase and glucoamylase [8,9], or by combination of acid hydrolysis and enzymatic hydrolysis [7]. Hydrolysis is a chemical decomposition using water to separate the chemical bonds of the substance. Hydrolysis of starch is a breaking process of starch molecule to become constituent parts of the starch like dextrin, isomaltose, maltose, and glucose [10].

This experiment aimed to produced glucose syrup from red sorghum starch by hydrolysis enzymatic method process in various enzyme and substrate concentration.

2. Experimental details

2.1. Material

Red sorghum starch from Bandung was used in this study. It was grinded and sieved to obtain particle size of 125 μm. The component of red sorghum starch is listed in Table 1. Amylose and amylopectin in red sorghum starch can be converted into glucose syrup or liquid sugar.

Table 1. Red sorghum starch content.

Component	Amount (%)
Starch	52.7
- Amylose	18.6
- Amylopectin	34.1
Protein	8.9
Fiber	9.8

Table 2 show the characteristics of enzymes used in this experiment.

Table 2. Enzyme characteristics.

No	Enzyme characteristics	Enzyme	
		α-amylase	Glucoamylase
1	Form	liquid	liquid
2	Enzyme activity	148,39(KNU/g)	293(AGU/g)
3	Density (g/ml)	1,244 g	1,155
4	Total viable count (/g)	< 100	< 200
5	Coliform bacteria (/g)	< 4	< 4

2.2. Experimental

2.2.1. Enzymatic hydrolysis. Red sorghum starch as the substrate was diluted at 10, 20, 30, and 40% for each 300 ml substrate solution (w/v), while the enzyme concentration was varied at 0.2, 0.27, and 0.30% (v/v) for each substrate concentration. In gelatinization, substrate solution was stirred and heated until 60°C then α-amylase was added and continued for liquefaction. The temperature of the solution was increased until 90°C for 60 min. Samples were taken every 10 min for brix value determination. Saccharification was carried out by decreasing the temperature to 60°C. Glucoamylase was added at the ratio 1:1 to α-amylase after liquefaction. The heating process was set in a constant temperature at 60°C for 120 min with sampling rate every 10 min for brix value as the total sugar concentration. All steps of the experiments were observed in constant pH 6.

2.2.2. Reducing sugar and glucose content. Reducing sugar concentration was determined A 3,5-dinitrosalicylic acid (DNS) solution and the samples absorbance were analyzed by spectrophotometer (Labo) with visible detector and wavelength of 540 nm [11].

High liquid performance chromatography (HPLC) was used to observed the glucose and fructose contents. HPLC detector used was Agilent 1260 refractive Index Detector with positive polarity; Agilent Zorbax Carbohydrate 4.6 x 150 mm, 5 μ column was used in this observation.

3. Result and Discussion

Enzymatic hydrolysis was carried out in three steps namely gelatinization, liquefaction, and saccharification. Gelatinization is an important step where starch grains are heated with excess of water to increase amylopectin amorphous region and enzyme accessibility. Gelatinization is the prefix process before liquefaction. It is a process of swelling of starch granules due to heating which breaks the hydrogen bond on the starch glycoside bond. The swelling of the granule is irreversible or cannot be back to its original shape. Liquefaction carried out without gelatinization will take longer time than the gelatinized substrate [12].

3.1. Liquefaction Process

Liquefaction was carried out by amylases that hydrolyzed the chemical bond α-(1-4) of starch, producing dextrin, maltose, maltotriose and maltopentoses with a dextrose equivalent (DE) below 30 at high temperature [13]. The result of liquefaction is shown in Figure 1a-d. The total sugar concentration was measured in terms of brix value for each process.

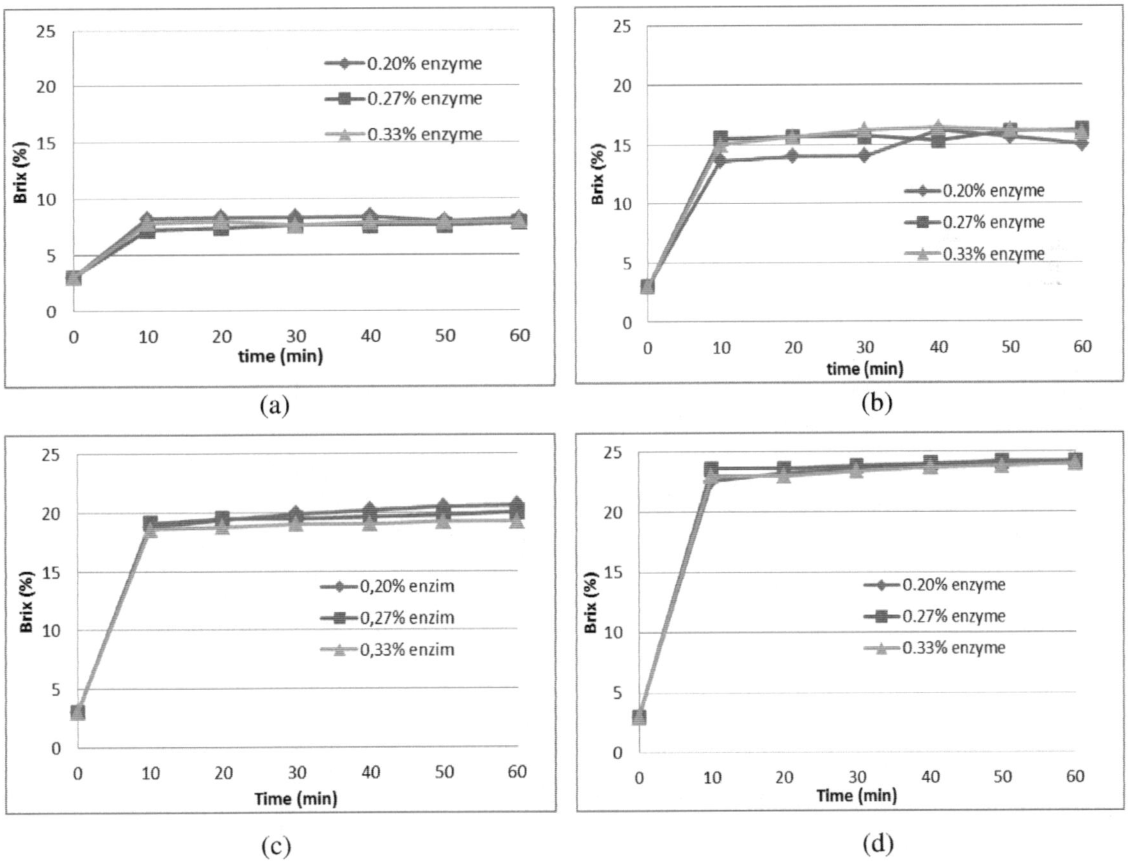

Figure 1. Total sugar level in liquefaction process with varied substrate and enzyme concentration (a) 10% substrate, (b) 20% substrate, (c) 30% substrate, (d) 40% substrate.

All figures show that total sugar increased rapidly in the first 10 min and slowly for the next run time. It indicated that α-amylase could convert the red sorghum starch rapidly because of the high enzyme activity. Increasing of total sugar concentration that showed in % brix was similar with increasing of reducing sugar concentration. Increasing of reducing sugar concentration indicated that the product was formed as a function of time for a series of substrate concentrations. As expected, the amount of product formed increased with time, although eventually a time was reached when there was no net change in the concentration of substrate or product. By the Michaelis-Menten Model, the enzyme was still actively converting substrate into product and vice versa, but the reaction equilibrium had been attained [14]. In various concentration of additional enzyme, the brix value increased rapidly in the first 10 min then it rose up slowly until constant brix value above 23%. It indicated that the amylose and amylopectin were running out at those substrate solutions become the product. For the 10% substrate concentration showed the maximum total sugar formed until brix 8.2%. It was the lowest substrate concentration so the amylose and amylopectin which were converted to glucose were lowest than the other concentration of substrate.

3.2. Saccharification

Saccharification process of partially hydrolyzed starch was carried out at lower temperatures. Saccharification by glucoamylase to hydrolyze the chemical bonds α-(1-4) and α-(1-6), produces maltose or D-glucose syrups with DE around 40 and 96, respectively [13].

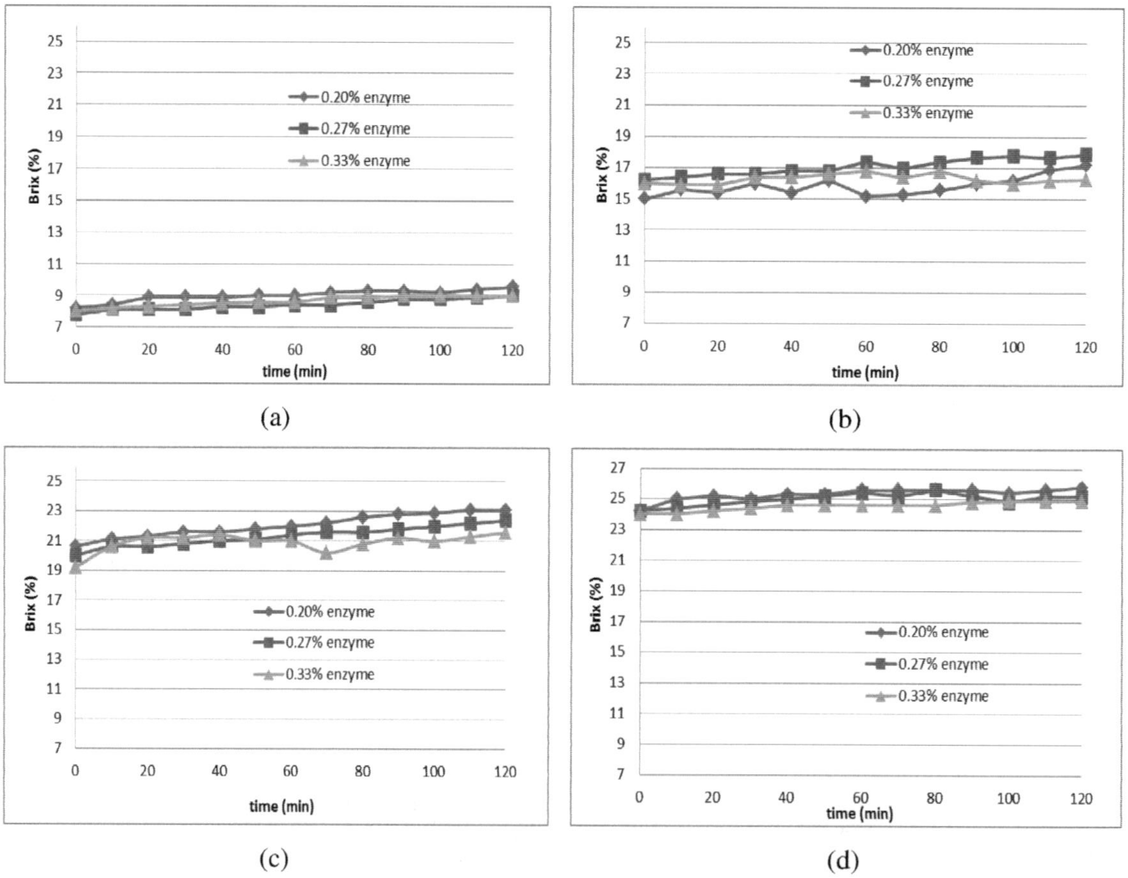

Figure 2. Total sugar level in saccharification process with varied substrate and enzyme concentration (a) 10% substrate, (b) 20% substrate, (c) 30% substrate, (d) 40% substrate.

The result of saccharification is shown in Figure 2a-d. The brix value in those figures increased slowly and became constant. It means that the total sugar production had almost reached the maximum value in the end of liquefaction process for each substrate concentration. The addition of glucoamylase in the solution did not provide a significant improvement to the brix value. The products were then analyzed for the reducing sugar as the glucose concentration. Glucoamylase converted the dextrin to produce D-Glucose or maltose [15]. It would appear if the solution was analyzed with DNS reagent as the reducing sugar concentration.

3.3. *Identification of hydrolysis end product*

The end product of each variation of condition is shown in Figure 3. The reducing sugars increased slowly as substrate concentration increasing at every enzyme concentration. This means that the reaction rate did not depend on the substrate concentration [16]. The addition of enzymes decreased starch in the substrate. The higher enzymes concentration in the solution, the higher the binding of substrate to the active side of the enzyme. When the substrate was entirely converted into the product, the increase of enzyme concentration showed no increase of product amount [17]. The reducing sugar as the end product become the accumulation of sugar formed as long as the liquefaction and saccharification occurred.

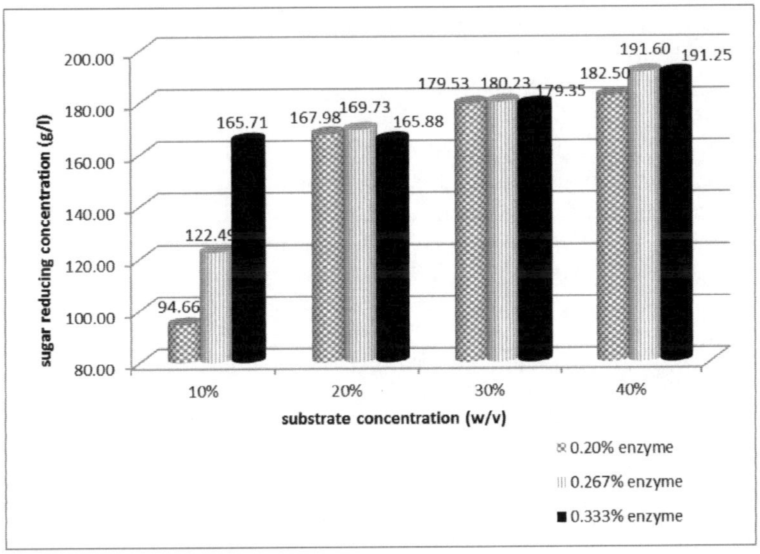

Figure 3. Concentration of hydrolysis product for varied substrate and enzyme concentration.

Figure 3 shows that the highest reducing sugar concentration was made from 40% substrate at any various enzyme concentration. The reducing sugar concentration were 182.50 g/l, 191.60 g/l, and 191.25 g/l respectively for each enzyme concentration. It is still possible to enhance the reducing sugar concentration by increasing the substrate concentration. However, this addition caused the solution too viscous [15]. Therefore, it was difficult to hydrolyze by acidic or enzymatic treatment.

The glucose formed was of 17.44% for the highest end product concentration as of 40% substrate concentration, 0.27% enzyme concentration.

4. Conclusion

Enzyme and substrate concentration play a role in the production of glucose syrup from red sorghum starch. The best result was obtained for 40% substrate and 0.267% enzyme concentration with 191.60 g/l reducing sugar and 17.44% glucose.

Acknowledgments
We thankful to Ditjen Penguatan Riset dan Pengembangan (DRPM) Ministry of Research, Technology, and Higher Education for their financial support, UPPM Politeknik Negeri Bandung, and the Bioprocess Laboratory in Department of Chemical Engineering for their support.

References
[1] Ta'awu K G, Gernah D I and Igbabul B D 2014 *Brit. J. Appl. Sc. Technol.* **4** 892-904
[2] Das S, Bhattacharya A, Haldar S, Ganguly A, Sai Gu, Ting Y P and Chatterjee P K 2015 *Sust. Mat. Technol.* **3** 17-28
[3] Yadav P and Majumder C B 2017 *J. Interg. Sci. Technol.* **5** 19-22
[4] Silva R N, Quintino F P, Monteiro V N and Asquieri E R 2010 *Cienc. Tecnol. Aliment.* **30** 213-7
[5] Singh H, Singh S N and Singh N 2009 *Int. J. Food Properties* **12** 713–25
[6] Udachan, S Iranna, A K Sahoo and G M Hend 2012 *Int. Food Res. J.* **19** 315-9
[7] Triyono A 2008 *Pros. Sem. Nas. Teknoin 2008 Bidang Teknik Kimia dan Tekstil* B7-10
[8] Ruiz M I, Shancez C I, Torres R G and Molina D R 2011 *J. Braz. Chem. Soc.* **22** 2337-43
[9] Permanasari A R and Yulistiani F 2015 *J. Fluida Tek. Kimia Polban* **11** 9-14
[10] Terahara N, Konczak I, Ono H, Yoshimoto M and Yamakawa O 2004 *J. Biomed. Biotechnol.* **5** 279–86
[11] Adney B and Baker J 1996 *Measurement of Cellulase Activities Laboratory Analytical Procedure (LAP)* Technical Report NREL/TP-510-42628
[12] Mitsuiki S, Mukaea K, Sakai M, Goto M, Hayashida S and Furukawa K 2005 *Enzyme Microb. Technol.* **37** 410-6
[13] Kunamneni A, Permaul K and Singh S 2005 *J. Biosci. Bioeng.* **100** 168-71
[14] Berg J M, Tymoczko J L and Stryer L 2002 *Biochemistry 5th ed.* (New York: W H Freeman)
[15] Permanasari A R, Yulistiani F and Djenar N S 2017 *Adv. Sci. Lett.* **23** 5775-9
[16] Lee J M 2009 *Biochemical Engineering* (New Jersey: Prentice-Hall Inc. 1992)
[17] Poedjiadi A 1994 *Dasar-dasar Biokimia (Edisi Revisi)* (Jakarta: UI Press)

Physical and chemical properties of corn, cassava, and potato starchs

A H D Abdullah[1*], S Chalimah[1], I Primadona[1] and M H G Hanantyo[2]

[1] Research Unit for Clean Technology, Indonesian Institute of Sciences, Bandung, Indonesia
[2] Program Study of Material Engineering, Faculty of Mechanical and Aeronautical Engineering, Institute Technology of Bandung, Bandung, Indonesia

[*]Email: dawamdullah@yahoo.com

Abstract. Starch-based plastic is developed for petroleum-based plastic material replacement. Thus, basic knowledge of starch properties is important. This research aimed to evaluate the physical and thermal properties of various potential starch, i.e. corn, cassava, and potato starch. Granule size, thermal property, and functional group of starch were determined by optical microsopy, DSC, and FTIR, respectively. Our results demonstrated that the properties of starch (e.g. granule morphology and thermal property) varied according to its different sources. Potato starch has a bigger granule size and lower gelatinization temperature compare to corn and cassava starch. This implied that the granule size of starch affected gelatinization temperature. The larger the granule size, the more easily the starch to be gelatinized.

1. Introduction

Plastic consumption in Indonesia in 2014 is about 4.2 million tonnes or approximately 10 kg per capita per year [1]. Indonesia Plastic Association (Inaplast) estimates that these numbers will increase up to 7 million tonnes by 2018 [1]. Meanwhile, more than 99% of plastic raw materials are still dominated by petroleum [2]. Several commonly used plastics such as polyethylene (PE), polypropylene (PP), low density polyethylene (LDPE), high density polyethylene (HDPE), and polyvinylchloride (PVC) have caused abundant of waste which cannot be decomposed. It will accumulate and causes environmental pollution [3,4]. In addition, due to the non-renewable resources, these plastic materials are constantly diminished in the nature [5]. In terms of environmental and sustainability issues, therefore, it is urgent to develop biodegradable plastic as a substitute for non-biodegradable conventional plastic.

Starch as a common polysaccharide is a polymer made of large number of saccharide molecules and potential to be developed as biodegradable plastic [6-10]. Starch is available in nature from various types of commodities such as corn, cassava, and potato. Structure of starch is a semi-crystalline granular shaped molecule composed of two types polysaccharides namely amylose and amylopectin. Amylose has relatively straight chain structure, whereas amylopectin is a polysaccharide molecule that has a branched chain structure [10]. The size and shape of the starch granules is determined by the branched structure of amylopectin rather than amylose. While the strength of the starch granules is determined by the degree and value of the polydispersity index of amylose [11]. The ratio between amylose and amylopectin in every starch's granule depends on the type of starch. However, in general, starch contains 75-80% amylopectin. The amount of amylose is smaller than amylopectin due to molecular size of amylose which is smaller than amylopectin [12].

IOP Conf. Series: Earth and Environmental Science **160** (2018) 012003 doi:10.1088/1755-1315/160/1/012003

The small granular size of amylose and amylopectin allows them to link with protein matrix, and changes their thermal properties such as gelatinization temperature. This will lead to reduce the quality of plastics made [13]. The size and shape of granular particles will vary depending on the source and part of the plant. Furthermore, plants with the same variety can produce different sizes and shapes of granules depending on the growing conditions of the plant [14].

In the manufacture, the size of the granules will affect the bioplastic properties. The larger the granule, the thicker the bioplastic could be made, and further lowering strength, elongation, and tensile strength [15]. Herein, we report the granule morphology, thermal property, and a range of functional groups of three types of potential local starches (i.e. corn, cassava and potato). We have shown that these three starches exhibited different morphology structures and gelatinization temperatures.

2. Method

2.1. Materials
All chemicals and reagents were used as received. Corn, cassava, and potato starch were purchased from PT. Markaindo Selaras, PT. Budi Starch, and PT. Makmur Food, respectively.

2.2. Starch characterization
The size distribution of starch granules was examined using Optical Microscope (OM) Type Meiji Techno MT6000 combined with ImageJ software. About 1 mg of starch powder was fixed on microscope glass holder. The thickness of powder layer was made as thin as possible (< 0,1 mm) for clearly observation. The starch granules were observed by using a white light of optical microscope. The obtained data was then processed by ImageJ software to get the granule size distribution.

The gelatinization temperature was determined using Differential Scanning Calorimetry (DSC, NETSCH DSC 214). The sample was suspended in distilled water with mass ratio 1:1 (starch:H_2O) and then put into the aluminum crucible as much as ~5 mg for DSC experiment. The temperature range of DSC experiment was 30-100 °C with the heating rate 5 °C/min.

To evaluate and identify the functional group of starch, the Fourier Transform Infrared (FTIR, Shimadzu IR Prestige-21) technique was utilized. In this experiment, the starch powder was used as received.

3. Results and Discussion

3.1. Morphology of starch's granule
The OM image, which is presented in Figure 1, reveals that each starch possess a different granule shape and size. However, the surface of the corn and cassava granules (Figure 1a and b) are almost similar which tended to separate from each other and had the shape of oval - angular - spherical. Meanwhile, the shape of the potato granules tend to be ellipsoid and oval (Figure 1c). The analysis by using Image J software resulted distribution and average of granule size of corn, cassava and potato starches are 15.9, 15.3, and 26.9 µm, respectively ((Figure 2). As shown in Figure 2a and b, the size distributions of corn and cassava granules were medium and narrow distribution. Whereas, potato granules were bimodal with two general distribution namely medium and large distribution at 8-40 and 50 µm, respectively. The overall results of this experiment inform that the smallest granular size was owned by cassava starch, followed sequentially by corn and potato starch.

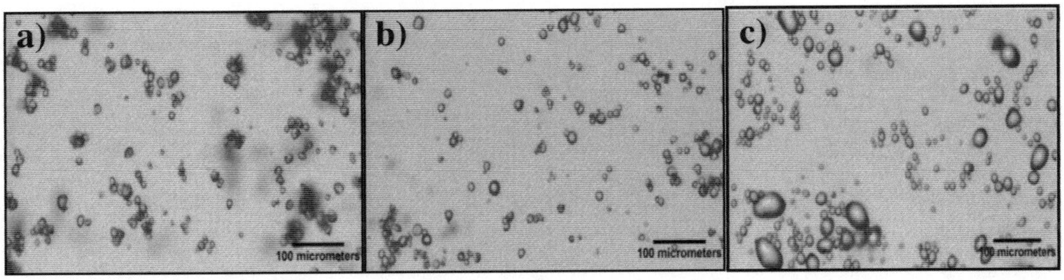

Figure 1. OM image of various types of starch: a) corn starch, b) cassava starch, and c) potato starch.

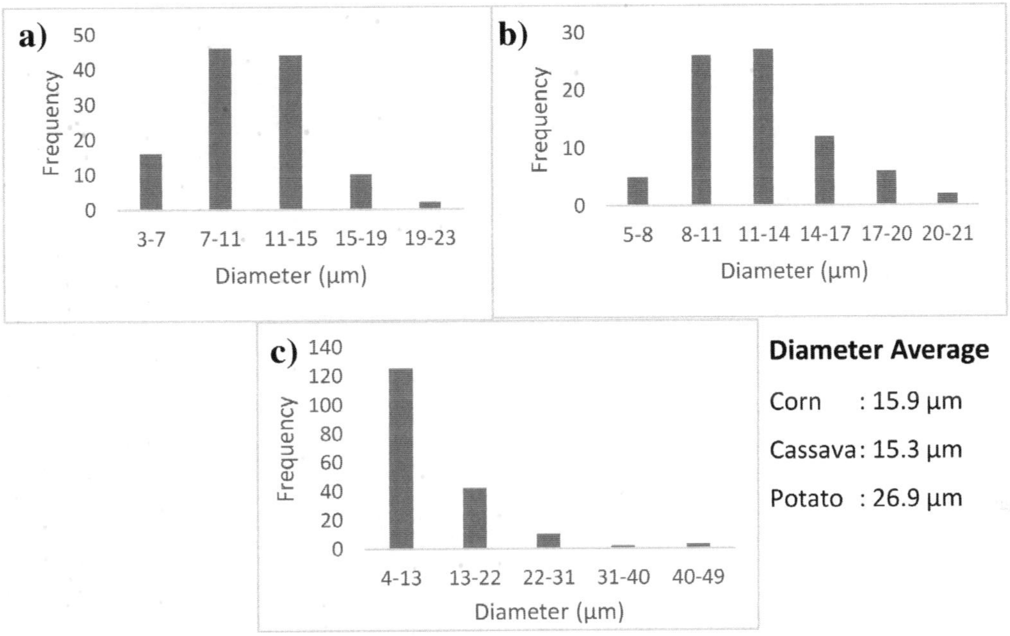

Figure 2. Distribution of granule size of various types of starch: a) corn starch, b) cassava starch, and c) potato starch.

3.2. Thermal properties of various types of starch

Gelatinization of starch is indicated by the thermal property as shown in Figure 3. The lowest gelatinization temperature is shown by potato starch, approximately 62 °C, whereas corn and cassava starch at about 66 and 68 °C, respectively. Higher of temperature gelatinization derived from a higher degree of crystallinity, which provides thermal stability of granule against gelatinization. Crystallinity is influenced by chain length of amylopectin [16,17]. The smaller the starch granule, the higher the gelatinization temperature.

In the bioplastic production, the gelatinization temperature of starch is crucial. If the gelatinization temperature is too low ($\leq$ 60 °C), the starch slurry might be gelatinized in the drying process and this condition is undesirable. Conversely, if the gelatinization temperature is too high, it causes the high energy consumption in the bioplastic production.

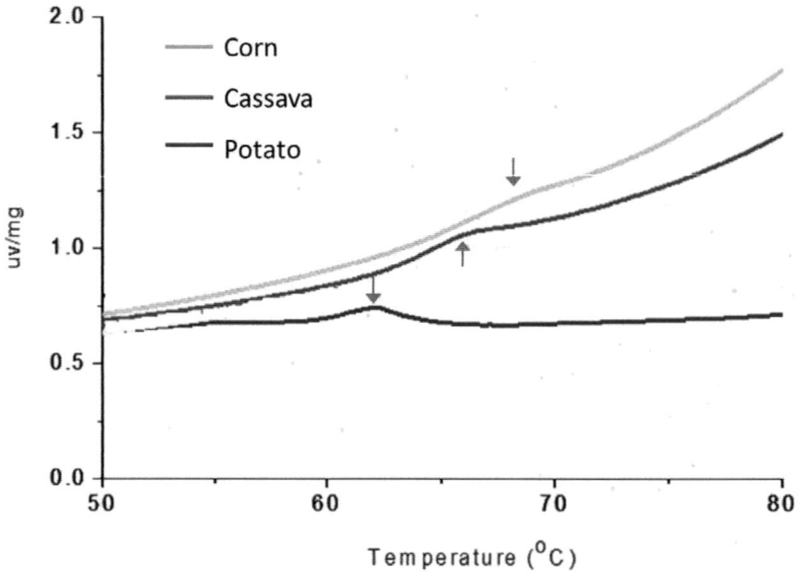

Figure 3. DSC spectrum of corn, cassava, and potato starch. Red arrow denotes the gelatinization temperature.

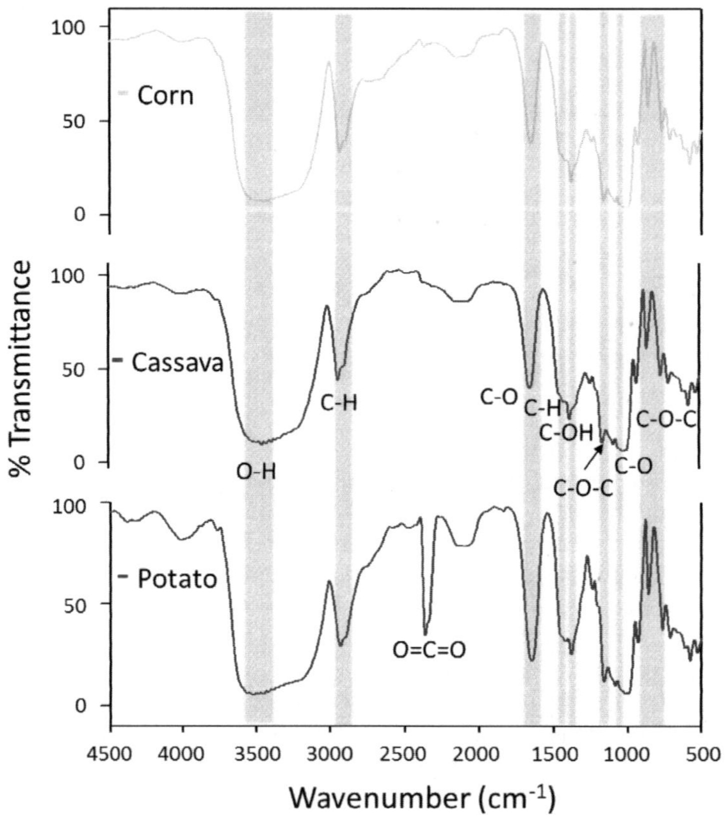

Figure 4. FTIR spectra of corn, cassava, and potato starches.

Table 1. Band assignment of corn, cassava, and potato starches.

No	Functional groups	Wave number literature	Corn	Cassava	Potato
1	O - H stretching	3600 – 3300	3448	3448	3523
2	C - H stretching	2931	2929	2930	2927
3	C-O bending associated with OH group	1637	1647	1646	1645
4	CH_2 symmetric deformation	1458	1437	1437	1437
5	CH_2 symmetric scissoring	1415	1415	1417	1419
6	C -H symmetric bending	1385-1375	1381	1381	1381
7	C - O - C asymmetric stretching	1149	1157	1157	1157
8	C - O stretching	1200 – 800	1082, 1016	1082, 1018	1082, 993
9	C - O - C ring vibration of carbohydrate	920, 856, 758	929,860, 763	929, 860, 763	929, 858, 763
10	O = C = O	2350	-	-	2358

3.3. FTIR analysis

The FTIR spectra of starches are presented in Figure 4 whereas the interpretation of each peak is given in Table 1. As shown in the Figure 4, the presence of absorption band at around 3300-3600, ~2900, ~1150, and 1000-1100 cm^{-1} in the three spectra indicated that all starches possess an OH, C-H, C-O-C, and C-O functional group, respectively. In addition, the characteristic C-O-C ring vibration on starch lead to an absorbance peak at around 700-900 cm^{-1}. The C-O bending associated with the OH group would cause an absorbance peak at around 1648 cm^{-1}. Furthermore, the absorbance peak at 1415 cm^{-1} implied the presence of C-H symmetrical scissoring of CH_2OH moiety. The uncommon CO_2 peak in starch (λ 2358 cm-1) was observed in the potato starch IR spectrum (Figure 4). Its might be results from measuring conditions. From this FTIR analysis, it shows that starch of corn, cassava, and potato possess similar chemical structure.

4. Conclusion

We have demonstrated here some physical and chemical properties of three types of corn, cassava, and potato starch). Those starches have a similar chemical structure. However, compared with cassava and corn, the granule size of potato is the biggest and possesed the lowest gelatinization temperature. The granule size appeared to correlate with gelatinization temperature. The smaller the size of starch granule, the higher the gelatinization temperature. This study provides additional information for bioplastic development in the future.

Acknowledgement

We would like to thank Kemenristek Dikti for the financial support for this research through the Insinas 2017 program.

References

[1] Global Business Guide Indonesia 2016 *Indonesia's Plastik and Packaging Industry Still Dependent on Raw Material Imports* http://www.gbgindonesia.com/en/manufacturing/article/2016/indonesia_s_plastik_andamp_packaging_industry_still_dependent_on_raw_material_imports_11500.php

[2] Sprajcar M, Horvat P and Krzan A 2012 *Biopolymers and BioPlastik* (Ljubljana: National Institute of Chemistry)

[3] Tanase E E, Rapa M and Popa O 2014 *Sci. Bull. Ser. F Biotechnol.* **18** 188-95

[4] Kolybaba M, Tabil L G, Panigrahi S, Crerar W J, Powell T and Wang B 2003 *ASAE Paper* RRV03-0007 (St. Joseph, Mich., USA: The Society for Engineering in Agricultural, Food, Biological Systems)
[5] Avella M, De Vlieger J J, Errico M E, Fischer S, Vacca P and Volpe M G 2005 *Food Chem.* **93** 467-74
[6] Elgadir A M, Akanda M J, Ferdosh S, Mehrnous A, Karim A A, Noda T and Sarker M Z 2012 *Molecules* **17** 584-97
[7] Lu D R, Xiao C M and Xu S J 2009 *eXPRESS Poly. Lett.* **3** 366-75
[8] Fang J and Fowler P 2003 *J. Food Agric. Environ.* **1** 82-4
[9] You C, Chen H, Myung S, Sathitsuksanoh N, Ma H, Zhang X Z, Li J and Zhang Y H P 2013 *PNAS* **110** 7182-7
[10] Solomon T W G, Fryhle C B and Snyder S A 2013 *Organic Chemistry 11th ed*. (USA: John Wiley & Sons, Inc.)
[11] Gallant D J, Bouchet B and Baldwin P M 1997 *Carbohydr. Polym.* **32** 177-91
[12] Brown W H and Poon T 2005 *Introduction to Organic Chemistry (3rd ed.)* (USA: John Wiley & Sons, Inc.)
[13] Cornell H J, Hovering A W, Chryss A and Rogers M 1994 *Starch* **46** 203-7
[14] Daniel J R, and Whistler R L 1990 *Cereal Foods World* **35** 825
[15] Lim S, Jane J, Rajagopalan S and Seib P A 1992 *Biotechnol. Prog.* **8** 51-7
[16] Singh N, Singh J, Kaur L, Sodhi N S and Gill B S 2003 *Food Chem.* **81** 219-31
[17] Singh S, Singh N, Isono N and Noda T 2010 *J. Agric. Food Chem.* **58** 1180-8

Antibacterial activity of bacterial cellulose-based edible film incorporated with *Citrus* spp essential oil

Y Srikandace*, L Indrarti, Indriyati and M K Sancoyorini

Research Unit for Clean Technology, Indonesian Institute of Sciences (LIPI)
Komplek LIPI Bandung, Jl. Sangkuriang, Gedung 50, Bandung 40135, Indonesia

*E-mail: (yoice_srikandace@yahoo.co.id)

Abstract. Bacterial cellulose-based edible films have been known as natural material for packaging containing carboxymethyl cellulose (CMC) and glycerol. Edible film is also known as an external food protection since it contains antibacterial substance. Incorporating antimicrobial compounds with essential oil (EO) from *Citrus* spp such as lime, lemon and sweet orange into edible films provides a novel approach to improve the safety, shelf-life of foods or fruits as well as physical properties of the edible film. The present study was aimed to investigate the antibacterial activity of edible film incorporated with EO. Antibacterial activity performance was carried out using broth and dilution agar method to determine EO role to inhibit the growth of pathogen bacteria such as *Escherichia coli* and *Staphylococcus aureus*. The results showed that edible film containing 2% lime, lemon and sweet orange performed moderate antibacterial activity with minimum inhibition concentrations (MICs) of about 100, 250 and 225 mg/mL, respectively.

1. Introduction

Biocellulose or microbial cellulose is one of natural and eco-friendly fibers and have important economic globally. Bacterial cellulose fibers are very pure, very stiff and similar to plant cellulose. Commonly, biocellulose is produced by *Acetobacter* species, displays unique properties including high tensile strength, high water absorption capacity, high crystallinity and an ultra-fine as well as highly pure fiber nanofibrillar structure [1,2]. Biocellulose has a diameter of 20-50 nanometers—1000 times thinner than the average human hair. The fiber length are interconnected and well-arranged, resulting in forming a network structure [2,3]. Nowadays, bacterial cellulose as a biological nonwoven fabric with special emphasis to provide the application of natural and biodegradable material package in order to reduce the use of synthetic materials that contribute to environmental contamination has been studied by many researches [4]. Biocellulose is now used for beauty treatments as well, since biocellulose composite has been introduced to enhance biocellulose properties through addition of reinforcement materials [3].

Biocellulose is grown in laboratory by tiny bacteria Acetobacter that convert glucose into cellulose and need oxygen to survive which resulted in forming the fibers. *Acetobacter xylinum* is a gram negative rod-shaped microorganism which is also known as *Gluconacetobacter xylinus*. The usage of glucose as carbon source during biocellulose production is associated with the formation of gluconic acid in the culture medium that affect the quantity of biocellulose production [2,3]. Edible film from nata de coco is one of the products generated from *Acetobacter* spp.

Edible films have been used for a long time for packing food. Currently, they are undergoing an advanced development [5,6]. Film or coating is a thin layer of material used for wrapping or coating layers among the foods. Currently, materials of film manufacturing are bio-based, naturally biodegradable and environmental friendly. These newer film materials have replaced conventional plastic films because of their harm to the environment [1]. The natural biodegradable films are normally applied in food products in order to prevent their spoilage and extend their shelf life [7].

Edible films also have been used as carriers of foods additives such as antioxidant or antimicrobial agent to improve the characteristic of the edible film and to reduce the materials that contribute to environmental contamination [4]. One of antimicrobial agents are essential oils (EOs) which demonstrated their ability in inhibiting microbial population in order to help food products with higher quality and safety [5,8]. Initial experiments using plate overlay assays demonstrated that 2% oregano EO was active against *Staphylococcus aureus* and *Salmonella*. Two percent rosemary EO was active against *S. aureus*, *L. monocytogenes*, *E. coli* and *S. typhimurium* [9]. The study of the antimicrobial effect of edible films containing 1% of oregano oil and cinnamaldehyde was evaluated in strawberry and spinach during cold storage. These results indicated that edible films containing EOs and their constituents can be used as release system for volatile antimicrobial agents to improve the safety and shelf life of fresh-cut fruits and vegetables [6]. In recent study, EOs from *Citrus* spp (*Rutaceae*) are considered as semi-sweet taste material [10] and possesses as an antimicrobial [4]. Studies of incorporation of EOs into edible films to improve the microbiological safety of foods are still limited. This study aimed to determine the antibacterial activity of edible film incorporated with lime, lemon and sweet orange EO. The effects of EOs addition to the growth of pathogen bacteria according to the minimum inhibition concentration (MIC) were evaluated.

2. Methods

2.1.Materials and preparation of bacterial cellulose-based edible film

Lime, lemon and sweet orange oils were purchased from Lansida Group. Nata de coco gels were purchased from local industry in Cianjur, West Java Province, Indonesia. The washing, purification, and preparation of bacterial cellulose (BC) slurry were conducted as previously reported [10].

2.2. Preparation of bacterial cellulose composites

An amount of 200 mL of BC slurry was composted of 30% w/w glycerol (Merck), 30% w/w CMC (Himedia), and 1-2% essensial oil based on the gross weight of BC slurry. Each of additives was mixed with BC slurry in a beaker glass under magnetic stirring condition at 60°C. Subsequently, the solutions were degassed with vacuum bell jar to remove the bubbles, then poured on the tray. The trays were held overnight at 45°C in oven blower, then cooled at room temperature before peeling the films off the trays and stored in plastic bags until used [10].

2.3. Antibacterial activity of esensial oils and edible films

The antibacterial activity of essensial oils and films incorporated with EOs against the bacterial strains (*Escherichia coli* and *Staphylococcus aureus*) were tested and determined by the disc diffusion assay. Film samples were cut into discs with 6 mm in diameter using a sterile punch and placed on plates containing Mueller–Hinton agar (MHA) (Himedia) which had been previously seeded with 100 μL of an overnight broth culture containing approximately 10^5 CFU/mL of the test bacteria. The plates were incubated at 37°C for 24 h and the clear zone formed around the film disc on the media was recorded [6].

The minimum inhibition concentrations (MIC) of active edible films containing EOs were also determined. The MIC value was evaluated as the lowest concentration of samples. Three replicates were conducted for each microorganism. The plates were incubated under anaerobic conditions for 24 h in an oven maintained at 36 ± 1 °C and subjected to visual inspection. The presence of microbial

IOP Conf. Series: Earth and Environmental Science **160** (2018) 012004 doi:10.1088/1755-1315/160/1/012004

growth on the medium indicated that the essential oil possessed bacteriostatic activity, while the absence of the growth implied bactericidal activity of the oil sample [11,12].

3. Result and Discussion

3.1.Antibacterial activity of essensial oils
The growth inhibition zones of EOs were measured using agar disc diffusion assay. The antibacterial activity of essensial oils as antimicrobial agents are shown in Figure 1.

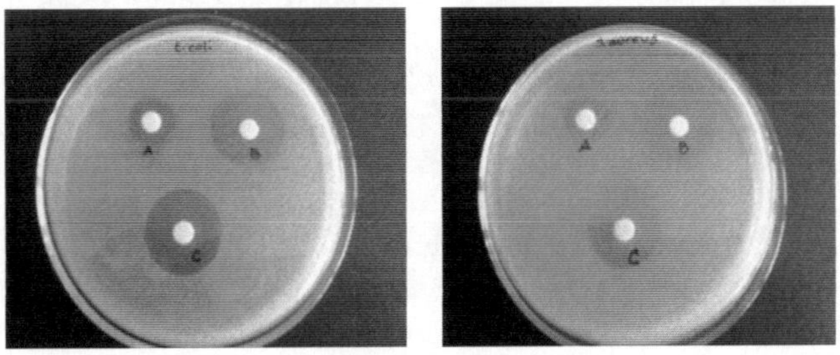

Figure 1. The inhibition zones of essential oils against pathogen bacteria *E. coli* (left) and *S. aureus* (right). A=Lime, B=Lemon and C=Sweet orange.

Figure 1 shows that the strong antibacterial activity for both bacteria was from lemon and sweet orange with inhibition diameter about 22 and 25 mm, respectively. Lime performed clear zone about 14 mm for both bacteria. This result reveals that both lemon and sweet indicates strong activities whereas lime as moderate. The current EO antimicrobial activity was suggested based on previous study of which bergamot EOs was characterized against *Listeria monocytogenes* [13]. In addition, it was also confirmed that sweet orange and lemon showed good antibacterial activity against both gram-negative and gram-positive bacteria. The MICs for selected EOs ranged 15–250 µg/mL [14].

The different ability as the antibacterial properties might be due to the chemical contents in EOs. Some studies showed that limonene and other metabolites which is belong to alkaloids, flavonoids, steroids, terpenoids, saponins, cardiac glycosides as active compound of *Citrus* spp could inhibit the growth of *E. coli*, *S. aureus*, *Salmonella thypiimurium* and *Bacillus cereus* cells [4,16,17]. Another study showed that the essential oils of cinnamon, oregano, thyme, and clove attribute to the existence of cinnamaldehyde, carvacrol, thymol, and eugenol, which is possess similar activities against *L. monocytogenes*, *Salmonella typhimurium*, *E. coli*, *Brochothrix thermosphacta*, and *Pseudomonas fluorescens* [18]. It indicated that bioactive compounds of essential oils possessed antibacterial properties against several pathogen bacteria.

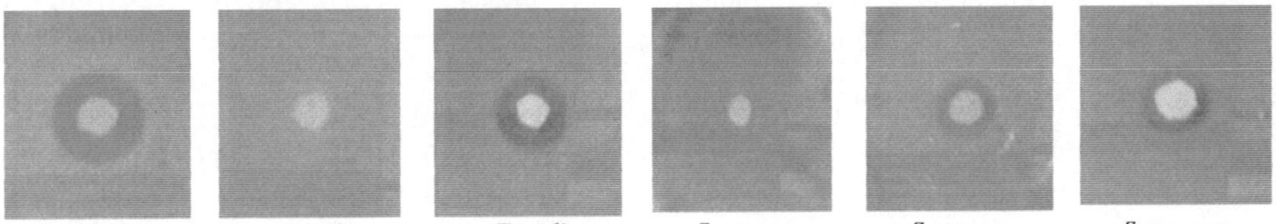

E. coli 2% lime	*E. coli* 2% lemon	*E. coli* 2% sweet orange	*S. aureus* 2% lime	*S. aureus* 2% lemon	*S. aureus* 2% sweet orange

Figure 2. The antibacterial activity of edible films each containing 2% lime, lemon and sweet orange.

3.2. Antibacterial activity of edible films

Edible films that were incorporated with 1-3% EOs also showed antibacterial activity (Figure 2). Addition of 1-3% EOs in edible films show the same clear inhibition zone. It might be due to the high saturation stage of EOs in BC slurry when it was gear oven-dried.

Figure 2 shows that the area of clear zones of edible films with EOs decreased compared with clear zones of disc containing EOs only (Figure 1). Zones of inhibitions for *E. coli* of the edible films containing lime, lemon and sweet orange were 18, 0.9, and 17 mm, respectively. On the other hand, diameter of inhibition for *S. aureus* of the edible film incorporated with lime, lemon and sweet orange were 0, 13, and 13 mm, respectively. Edible films containing EOs with clear zone of $\theta = 13$, 17, and 18 mm show moderate activity, $\theta = 0.9$ m was weak activity, $\theta = 0$ mm was none activity.

The edible film with 3% orange oil against *E. coli* and *S. enterica* showed no inhibition area [5]. A study showed that addition EOs into film might reduce the interaction between protein molecules and contribute to the formation of discontinuous and uncompact texture with pores structure, resulting in film thickness increased [19]. Therefore, the ability of EOs gradually faded. The decrease of inhibition area of bacteria from edible film might be due to the polarity EOs to be bounded with BC slurry which contained high polarity.

The minimum inhibition concentration (MIC) of 2% EOs in edible films inhibited the cells of bacteria are shown in Figure 3. The MIC of each edible films containing the same EOs revealed the same concentration for both *E. coli* and *S. aureus*.

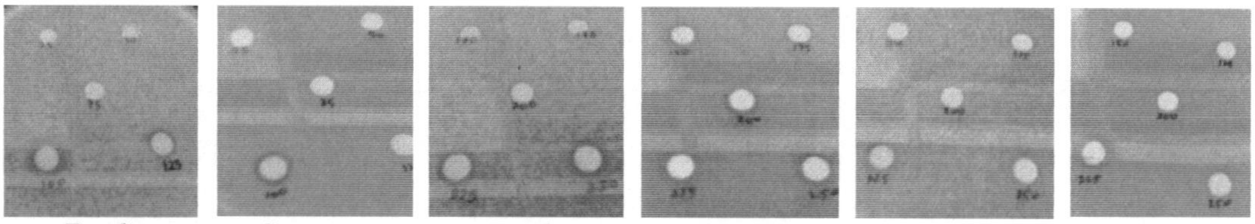

E. coli 2% lime | *S. aureus* 2% lime | *E. coli* 2% lemon | *S. aureus* 2% lemon | *E. coli* 2% sweet orange | *S. aureus* 2% sweet orange

Figure 3. Minimum inhibitory concentration (MIC) from edible films containing 2% EOs against *E. coli* and *S. aureus*.

Figure 3 shows the MICs of lime, lemon and sweet orange for *E. coli* and *S. aureus* are about 100, 250 and 225 mg/mL, respectively. Compared with previous study, of which 15.7 mg/mL oregano was effective against *E. coli*, *S. aureus,* and *L. monocytogenes* [20], current study required higher EOs concentration. This might be due to the carvacrol as the main compounds of oregano binds well lemon peel. Other results showed that 23 different citrus EOs were more effective against *S. Aureus*, *L. monocytogenes*, and *Salmonella enterica* as gram positive and gram bacteria, respectively [4].

The antibacterial potential for any edible films containing EOs depends on the physical, mechanical and chemical of material's characteristics. The present study showed that EOs from *Citrus* spp added to edible film could carry antibacterial agents in order to increase the value of edible film.

4. Conclusion

Based on the current study it is concluded that lime and sweet orange EOs were effective against *E. coli* and *S. aureus* with moderate activity whereas lemon was effective to inhibit *S. aureus*.

Acknowledgement

This research was funded by LIPI under the Riset Unggulan 2017 of which the authors acknowledged.

References

[1] Keshk S M A S 2014 *J. Bioproces. Biotechniq.* **4** 150
[2] Mohammad S M, Rahman N A, Khalil M S and Abdullah S R S 2014 *Adv. Biol. Res.* **8** 307-13

[3] Esa F, Tasirin S M and Rahman N A 2014 *Agric. Agric. Sci. Proc.* **2** 113–9
[4] Aldana D S, Andrade-Ochoa S, Aguilar C N, Contreras-Esquivel J C and Nevárez-Moorillón G V 2015 *Food Contr.* **50** 907–12
[5] Shaaban H A, Ali H S, Bareh G F, Al-Khalifa A R S and Amer M M 2017 *J. Appl. Sci.* **17** 171–83
[6] Du W-X, Avena-Bustillos R J, Hua S S T and McHugh T H 2011 *Science Against Microbial Pathogens: Communicating Current Research and Technological Advances* vol 2, ed A Mendez-Vilas (Badajoz: Formatex Research Center) pp 1124-35
[7] Valdés A, Ramos M, Beltrán A, Jiménez A and Garrigós M C 2017 *Coatings* **7** 56
[8] Matan N 2012 *Int. Food Res. J.* **19** 1733-8
[9] Morsy M K I, Khalaf H H, Sharoba A M, El-Tanahi H H and Cutter C N 2014 *J. Food Sci.* **79** 675-84
[10] Vasek O M, Caceres L M, Chamorro E R and Velasco G A 2015 *J. Nat. Prod.* **8** 16-26
[11] Crevelin E J, Caixeta S C, Dias H J, Groppo M, Cunha W R, Martins C H G and Crotti A E M 2015 *Evid. Based Complement. Alternat. Med.* **2015** 102317
[12] Radaelli M, da Silva B P, Weidlich L, Hoehne L, Flach A, da Costa L A M A and Ethur E M 2016 *Braz. J. Microbiol.* **47** 424–30
[13] Indrarti L, Indriyati, Syampurwadi A and Pujiastuti S 2016 *AIP Conf. Proc.* **1711** 050007
[14] Frassinetti S, Caltavuturo L, Cini M, Della Croce C M and Maserti B E 2011 *J. Essential Oil Res.* **23** 27-31
[15] Marotta S, Giarratana F, Parco A, Neri D, Ziino G, Giuffrida A and Panebianco A 2016 *Ital. J. Food Safety* **5** 210-3
[16] Oikeh E I, Omoregie E S, Oviasogie F E and Oriakhi K 2016 *Food Sci. Nutr.* **4** 103-10
[17] Saeb S, Amin M, Gooybari R S and Aghel N 2016 *Int. J. Enteric Pathogens* **4** e37103
[18] Mith H, Duré R, Delcenserie V, Zhiri A, Daube G and Clinquart A 2014 *Food Sci. Nutr.* **2** 403–16
[19] Wu J, Sun X, Guo X, Ge S and Zhang Q 2017 *Aquaculture Fish.* **2** 185-92
[20] Alvarez M V, Ortega-Ramirez L A, Gutierrez-Pacheco M M, Bernal-Mercado A T, Rodriguez-Garcia I, Gonzalez-Aguilar G A, Ponce A, Moreira M R, Roura S I and Ayala-Zavala J F 2014 *Front. Microbiol.* **5** 699

IOP Conf. Series: Earth and Environmental Science **160** (2018) 012005 doi:10.1088/1755-1315/160/1/012005

Isolation, identification of alkaloid from *Rhizophora mucronata* and the activity of its methanol extract against barnacles

D G S Andayani[1*], S R Anggraeni[2], E Liviawaty[2], R M Chrisentia[2] and Y Srikandace[1]

[1] Research Unit for Clean Technology, Indonesian Institute of Sciences (LIPI), Komplek LIPI Bandung, Jalan Sangkuriang, Gedung 50, Bandung, 40135, Indonesia
[2] Faculty of Fishery and Marine Science, Padjadjaran University, Jl. Raya Bandung Sumedang KM 21, Jatinangor 40600, Indonesia

[*]Email: desakgedesa@gmail.com

Abstract. A study was conducted to isolate, identify alkaloid content of *Rhizophora mucronata* and to determine its effectiveness of methanol extract against barnacles. The use of tributyltin (TBT) was effective to reduce biofouling. However, TBT has been prohibited because they are detrimental to non-target organisms and for surrounding environment. *Rhizophora mucronata* is useful as antifouling compounds because of the secondary metabolites. Its active compound will replace TBT as it is safe against the biota and the marine environment. The root's bark, trunk's bark and leaves powder were macerated using methanol. The methanol extract was used for total alkaloid isolation. The separation of alkaloids was done by preparative TLC and analyzed with FTIR and LC-MS. The results showed that the alkaloid metabolite identified was benzamide ($C_{17}H_{14}N_2O_2$). The methanol extract had antifouling activity against barnacles. The highest yield was found in *Rhizophora mucrophora* leaf extracts at about 24.8%. Antifouling activity test against barnacles showed that a mixture of wood paint and *Rhizophora mucronata* leaf, root and bark extracts were not significantly different.

1. Introduction

Submarine objects, especially in coastal waters, are attached by biofouling organisms. Barnacle is a macrofouling that forms a fouling community causing damage to beach buildings and increase fuel consumption, damage to pipes, pumps and vessels can be reversed in bad weather conditions [1,2]. Biofouling or biological fouling is the accumulation of microorganisms, plants, algae or animals on wetted surfaces [3].

Antifouling agent is a substance that prevents or retards fouling or marine underwater growth on plants, rocks or ships' bottoms [3]. Currently fouling prevention is done using antifoulant paint, which active compound is tributyltin (TBT). This compound is harmful to organisms. The copper content in TBT and continuous accumulation of TBT will poison the organisms in the sea, causing impaired immune systems and shells to form changes such as malformation [4,5].

The greatest interest in the use of natural products as antifoulants is based on their potential to function as nontoxic antifoulants. The new diterpene methoxy-ent-8 (14)-primarenely-15-one (1) and three metabolites: ent-8(14)-primarene-15R, 16-diol (2), stigmasterol (3) and β-sitosterol (4) were isolated from the roots of mangrove plant *Ceriops tagal* [6-8]. The objective of this research was to

IOP Conf. Series: Earth and Environmental Science **160** (2018) 012005 doi:10.1088/1755-1315/160/1/012005

isolate and identify alkaloids in *Rhizophora mucronata* extract derived from leaves, stems and roots, as well as to analyze the activity of the methanol extract as antifouling agent.

2. Methods

2.1. Extraction and fractionation

Isolation and fractionation of alkaloid metabolites was done from the roots, stems and leaves powder of mangrove. Extraction was carried out by maceration with methanol then concentrated using a rotary evaporator. The methanol extract was added with HCl 1 N, then extracted with ethyl acetate. The acid layer was added with ammonium hydroxide then re-extracted with ethyl acetate. The ethyl acetate layer was concentrated with a rotary evaporator to obtain a total alkaloid extract [9,10].

2.2. Separation, purification, and identification

The total alkaloid was purified by thin layer chromatography (TLC) and eluted with methanol, ethyl acetate, and hexane using silica gel plate 60 GF254. The spots were observed under UV lamp at 366 nm. Further separation was done using preparative TLC to obtain the pure alkaloid. The alkaloid isolate was analyzed and identified by FTIR and LC-MS [11,12].

2.3. Testing of alkaloid metabolite

The alkaloid of roots, barks and leaves of *Rhizophora mucronata* was determined with Dragendorff's test. Two mg of methanol extract was acidified with 2 M HCl and added by 1 mL of Dragendorff's reagent. Formation of orange or orange red precipitate indicated the presence of alkaloids. Wagner's test was done by preparing 2 mg of methanol extract, acidified with 1.5 % v/v of HCl, and added with several drops of Wagner's reagent. A yellow or brown color indicated the presence of alkaloids. Furthermore, the extract was added with HCl and NaOH followed by extraction with ethyl acetate.

2.4. Antifouling activity determination

The antifouling activity test was conducted in the sea waters of Teluk Awur Beach, Jepara, Central Java Indonesia. Temperature, pH, salinity, and brightness were recorded during the experiments for 16 days. Antifouling activity was observed by counting the number of macrofouling barnacle which attached on wooden board media. A 7 x 14 cm^2 area of wooden board was painted using a commercial paint mixed with extract of roots, stems or leaves of mangrove at 500 or 1000 ppm concentration. As a positive control used was wooden board of the same size painted with commercial antifouling, whereas the negative control used was the same commercial-painted wooden board without mixing with neither mangrove extract nor plain wood planks. The wooden boards were weighted and tied to a pivot on a dock and were spaced 50 cm from the sea level at the lowest tide; the distance between the boards and the beach was about 12 m [13-15].

2.5. Data analysis

The data was analyzed descriptively by standard deviation (STDEV) based on the number of attachment of macrofouling biota on wood board media. The standard deviation is a measure of how widely values are dispersed from the average value (the mean). The experiment was repeated three times. The standard deviation was calculated using the "n-1" method with the following formula:

$$\frac{\sqrt{\Sigma(x-\bar{x})2}}{(n-1)}$$

3. Results and discussion

3.1. Extraction and fractionation

Extraction by methanol submergence method was intended to bind the polar compounds. Alkaloids are polar compounds that have to be extracted with polar reagent. Methanol is the most polar reagent after water. The rendement of mangrove extract is shown in Table 1.

Table 1. Rendement of root's bark, stem's bark and leaves extracts of *Rhizophora mucronata.*

Sample	Sample weight (gr)	Extract weight (gr)	Rendement (%)
Root's bark	401.6	69.7	17.4
Stem's bark	500.1	89.2	17.9
Leaves	387.2	96.0	24.8

Rendement is a comparison between the weights of the ingredients used with the total weight of the material. The value of rendement is used to determine the effectiveness of bioactive metabolite material. The higher the yield value indicated the higher the value of the methanol extract produced. The highest rendement value was obtained from the leaves extract of *Rhizophora mucronata* because the leaves had smoother texture compared with stem's and root's barks. The tiny size of the leaves extract is easy to withdraw the compound. It correlates with the highest activity against macrobiota [7,8,11,16,17].

3.2. Alkaloid test of mangrove extract
The methanol extracts of roots, stems and leaves of *Rhizophora mucronata* showed positive alkaloid reactions proven by white deposits when Mayer and Dragendorff reagents were added (Table 2).

Table 2. Testing of Alkaloid metabolite of *Rhizophora mucronata.*

Alkaloid metabolite	Root's bark		Stem's bark		Leaves	
	Simplicia	Methanol extract	Simplicia	Methanol extract	Simplicia	Methanol extract
Sample + (Mayer + Dragendorff) reagent	+	+	+	+	+	+

+ = contains metabolite alkaloids

3.3. Separation and purification
The separation showed three spots of TLC. The ratio of eluent was 21 methanol: 6.5 ethyl acetate: 5.5 n-hexane. The results showed that the topmost spot of a reddish-purple shine indicated an alkaloid compound. The Rf spot was 0.9 (Figure 1).

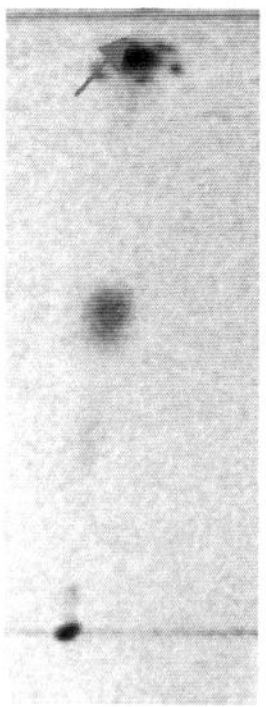

Figure 1. TLC of alkaloid isolate.

Separation of alkaloid compounds with preparative TLC was performed two yellow and one dark green bands. On the yellow bands are scraped subjected for purity by TLC again using the same developer.

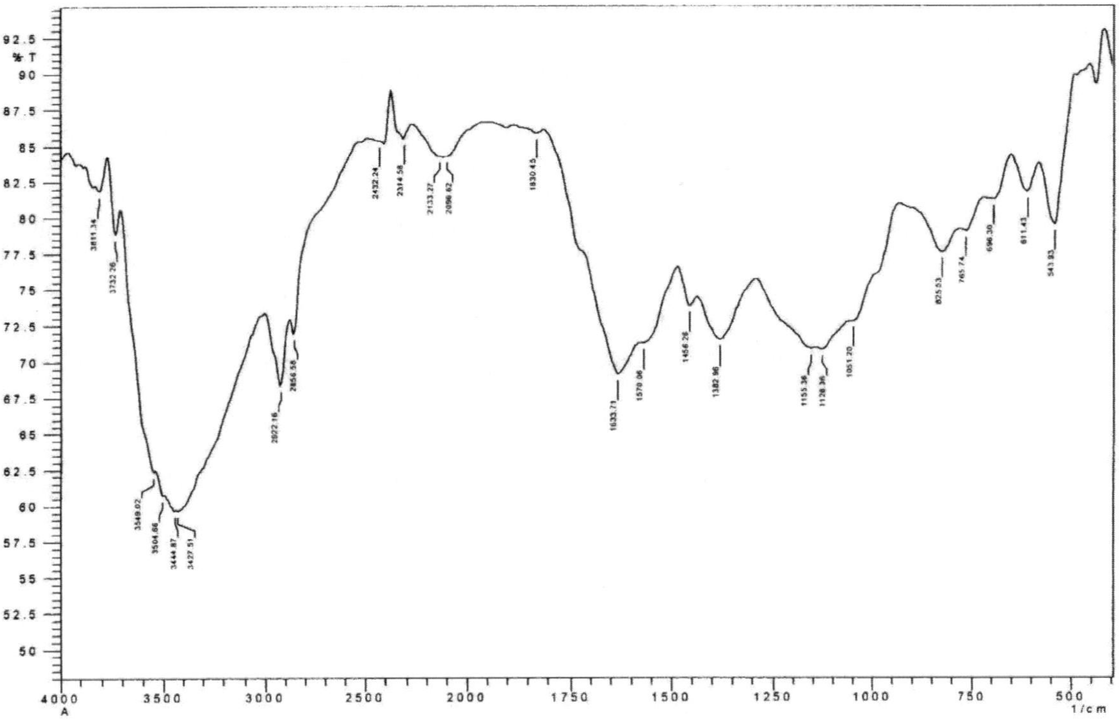

Figure 2. FTIR spectogram of alkaloids of mangrove leaves extract.

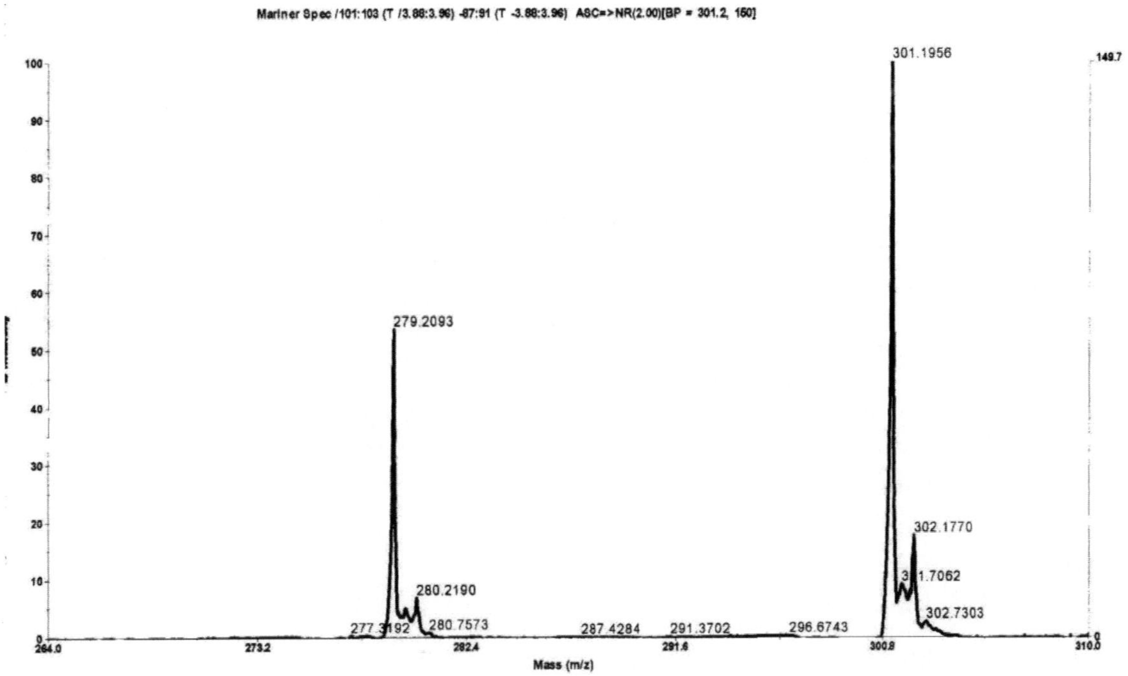

Figure 3. Spectogram of LC-MS.

3.4. Identification of compounds
Based on FTIR spectroscopy analysis, the alleged alkaloid compounds contained in the mangrove leaves were N-H, NH, C-O, C-N, CH and C-C groups. FTIR spectogram of alkaloids isolated from the leaves extract of *Rhizophora mucronata* is presented in Figure 2.

The results of FTIR spectroscopy showed the functional groups of the isolate samples. Figure 2 shows the vibration peaks with an absorption at the wavelength of 3444.87 cm^{-1}, which was the absorption of the vibration stretching of the NH group. It was also supported by the absorption at the wavelength of 1570.06 cm^{-1}, which was the vibration buckling of the NH group. The wavelength of 1051.20 cm^{-1} was a buckling vibration of C-O. The presence of vibration in the wave number 1382.96 cm^{-1} was the absorption of C-N, while at 2432.24 cm^{-1} wavelength was the bending vibration of CH and at 1633.71 cm^{-1} was the vibration of C-C group.

The isolate was analyzed using Liquid Chromatography-Mass Spectroscopy (LC-MS) to determine the molecular weight of the alkaloids. Spectogram of the LC-MS is shown in Figure 3.

The molecular weight of the isolated compound indicated by the LC-MS spectrometer was 278 g/mol. It is shown in Figure 3 in the presence of m/z 279.2093 [M + H]+ and m/z 301.1956 [M+Na]+. Based on these results, the isolate of alkaloids from mangrove leaf contained a benzamide compound, *N*-1-isoquinolinyl-3methoxy- with the chemical formula $C_{17}H_{14}N_2O_2$. The chemical structure of the benzamide alkaloid ($C_{17}H_{14}N_2O_2$) is shown in Figure 4.

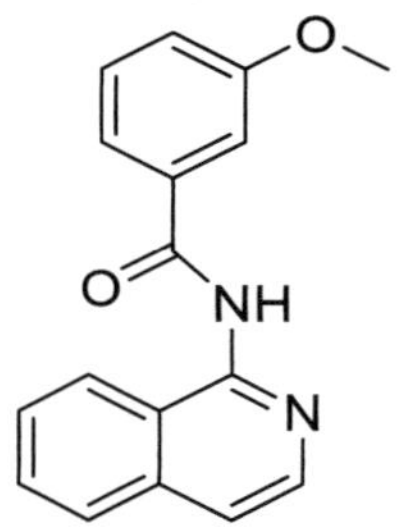

Figure 4. Alkaloid benzamide.

Six new 4-phenyl-3,4-dihydroquinolone derivatives along with the related of inquinolone 4 were isolated and identified from the cultures of an endophytic fungus obtained from the fresh leaves of the marine mangrove plant *Rhizophora stylose* [18]. Many bioactive metabolites including anthracenediones, xyloketals, sesquiter penads, chromones, lactones, coumarin and isocoumarin derivatives, xanthones and peroxides have been isolated from various mangrove-derived fungi in the South China Sea [19]. The methanol extract of *Rhizophora mucronata* contained alkaloid, flavonoid, and tannin [16], seven compounds were identified in the fraction of ethyl acetate of *Rhizophora mucronata* stem bark namely quinone, steroid, alkaloid and aromatic group with the highest secondary metabolite content was alkaloid group (74.8%) [20]. 2-isocyanato-2-hexanamine that was successfully isolated from mangrove plant was active against bacterial disease in shrimp [21]. The compounds contained in the skin of the mangrove stem *Sonneratia alba* is a phenolic group with lactone rings [22], the main secondary metabolites in mangrove plant in the form of alkaloid class of polar compounds. The white crystalline powder form characteristics had a melting point of 172°C with molecular weight of 232, and the molecular formula was $C_{12}H_{12}N_2O_3$ [23].

3.5. Antifouling activity
The data obtained was in the form of the number of attachment of macrofouling biota on wood board media. The data was analyzed descriptively with three replicates based on related sources in the form of journals, books, and thesis and presented in tabular form. The antifouling activity of *Rhizophora mucronata* extract at 500 and 1000 ppm was not significantly different with the amount of attachment between 247-315 barnacles (Table 3).

Table 3. Barnacles attachment.

Sample	Concentration of extract (500 ppm)			Concentration of extract (1000 ppm)		
Wood+paint+ extract (*Rhizophora mucronata*)	Root skin	Bark stem	Leaf	Root skin	Bark stem	Leaf
Barnacles attachment (individual)	280±35.68 a	315±15.95 a	253±42.74 a	282±23.67 a	286±47.51 a	247±66.16 a

Table 4. Barnacles attachment (control).

Control	Standard	Barnacles attachment (individual)
Positive (+)	Wood + commercial antifouling paint	4 ± 3.0a
Negative (-)	Wood + paint without antifouling	395 ± 60.68c
	Wood without painting	190 ± 98.38b

The results indicated that concentration of the extracts did not affect barnacles attachment. All values were not statistically different (Table 3).

3.6. The effectiveness of antifouling extract of Rhizophora mucronata

The bacteria initiate the attachment and form a primary thin film on the surface of submerged objects. The formed thin layer allows benthic diatoms, algae spores and larvae of various types of aquatic animals to attach and grow on the surface of the object. At the end, cauced bryozoa, hydrozoa, ciripedia, tunikata, algae and other biota to be a group of biota that developed later after forming layer film by bacteria. *Rhizophora mucronata* extract inhibited the growth of biofilm bacteria that reduced macrofouling attachment [17].

The number of barnacles attached to the wood media was due to the wood media surface was uneven and had many gaps so that the barnacles were attached to the cracks on the wooden surface. Selection of wood species for the test board was based on the type of wood commonly used in the manufacture of boats e.g. teak, meranti, and camphor. Meranti wood was selected because the price was more economical and easily obtained [24].

In addition to media and habitats, the attachment rate of barnacles was influenced by brightness, temperature, and current velocity [25]. Barnacles grow well at depth and brightness between 20-130 cm [26]. Water temperature during observation was 27°C on average; the temperature was quite good for the growth of barnacles because barnacles like waters with temperature of 15-35°C [26]. Soluble salinity or salinity in the study area was 36 ppt, the salinity was particularly suitable for growing barnacles because barnacles are able to live in estuary waters and have a salinity tolerance between 15-41 ppt [26]. The degree of acidity or pH of the waters at the time of observation was very good that reached 8, barnacles can live optimally at a pH between 6-9 [27].

Based on the number of wooden barnacles attached to wood with wood paint mixture, *Rhizophora mucronata* 1000 ppm leaf extract was a more effective anti fouling than root bark extract. However, the mixture of wood paint and *Rhizophora mucronata* 1000 ppm leaf extract would be much less effective when compared with commercial antifouling paint because the concentration of the active ingredients or secondary metabolites in commercial paints was very high compared with the active ingredients in mangrove leaf extract, due to the copper content.

4. Conclusion

Mangrove leaf extract (*Rhizophora mucronata*) contained betanidine alkaloid compounds ($C_{17}H_{14}N_2O_2$) and showed an antifouling activity against barnacles.

Acknowledgement
The authors would like to thank the management of Teluk Awur Beach, Jepara, Central Java Indonesia for allowing us to conduct antifouling activity test. The research was funded by Mandiri Research of Indonesian Institute of Sciences (LIPI).

References

[1] Fitridge I, Dempster T, Guenther J and de Nys R 2012 *Biofouling* **28** 649-69
[2] Gregory D B and Bharat B 2012 *Phil. Trans.R. Soc. A* **370** 2381-417
[3] Chambers L D, Stokes K R, Walsh F C and Wood R J K 2015 *Int. J. Env. Monit. Anal.* **3** 373-6
[4] Cao S, Wang J, Chen H and Chen D 2011 *Chinese Sci. Bull.* **56** 598-612
[5] Sudaryanto A 2001 *J. Tek. Lingkungan* **2** 241-6
[6] Chen J D, Feng D Q, Yang Z W, Wang Z C, Qiu Y and Lin Y M 2008 *Molecules* **13** 212-9
[7] Deepa S, Srikumar M and Padmakumar K 2014 *Int. J. Nat. Prod. Res.* **4** 54-7
[8] Cahyaningtyas G A, Iranawati F and Dewi C S U 2017 *J. Fish. Mar. Sci.* **1** 1-5
[9] Karin F 2010 *Encyclopedia of Life Sciences* (Chichester: John Wiley & Sons Ltd.)
[10] Vervoort H C, Richards-Gross S E, Fenical W, Lee A Y and Clardy J 1997 *J. Org. Chem.* **62** 1486-90
[11] Suganya R and Thangaraj M 2014 *Int. J. ChemTech Res.* **6** 4115-22
[12] Tims M 2017 *J. Nat. Prod.* **80** 1230
[13] Nandhini S and Revathi K 2016 *Int.J. Curr. Microbiol. App. Sci.* **5** 324-35
[14] Thummala V R R, Ivaturi M R and Nittala S R 2014 *Sci. Pharm.* **82** 247-63
[15] Raju C K, Pandey A K, Ghosh K, Pola A, Goud P S K, Jaywant M A and Navalgund S G 2017 *J. Pharm Biomed. Anal.* **140** 1-10
[16] Kasitowati R D, Yamindago A and Safitri M 2017 *J. Fish. Mar. Sci.* **1** 72-7
[17] Ernawati dan Hasmila I 2015 *J. Bionature* **16** 98-102
[18] An C Y, Li X M, Luo H, Li C C, Wang M H, Xu G M and Wang B G 2013 *J. Nat. Prod.* **76** 1896-901
[19] Wang X, Mau Z G, Song B B, Chen C H, Xiao W W, Hu B, Wang J W, Jiang X B, Zhu Y H and Wang H J 2013 *Mar. Drugs* **11** 3601-16
[20] Mahmiah, Sudjarwo G W and Hukmiyah M O M 2017 Seminar Nasional Kelautan XII Surabaya
[21] Suryati E, Rosmiati and Tenriulo A 2017 *Mar. Chim. Acta* **2** 19-23
[22] Herawati N 2011 *J. Chemica* **12** 54-8
[23] Darminto, Ali A, Dini I 2012 *J. Sainsmat* **1** 61-7
[24] Editor Regional Kompas 2009 http://regional.kompas.com/2009/03/31/20313832/perajin.perahu.nelayan.sulit.dapat.bahan.baku
[25] Nishizaki M T and Carrington E 2014 *J. Exp. Biol.* **217** 2101-09
[26] Nasmi J 2014 *Kelimpahan teritip (Balanomorpha) pada lambung kapal penunpang di Pelabuhan Kijang, Tanjung Pinang, Provinsi Kepulauan Riau* Bachelor Thesis (Bandung, Indonesia: Padjadjaran University)
[27] Campanati C, Yip S, Lane A and Thiyagarajan V 2016 *ICES J. Mar. Sci.* **73** 791-802

Phycobiliproteins production and heavy metals reduction ability of *Porphyridium* sp

S Priatni*[1], D Ratnaningrum[1], S Warya[2] and E Audina[2]

[1] Research Unit for Clean Technology, Indonesian Institute of Sciences (LIPI), Komplek LIPI Bandung, Jalan Sangkuriang, Gedung 50, Bandung, 40135, Indonesia
[2] School of Indonesian Pharmacy, Jl. Sukarno Hatta No 354, Bandung, Indonesia

*Email: sripriatni@yahoo.com

Abstract. *Porphyridium* sp is a unicellular red microalgae from Rhodophyta class which produces several high value polymers like polysaccharides, polyunsaturated fatty acids and phycobiliproteins. The phycobiliproteins contents are phycoerythrins, phycocyanins, and allophycocyanins. *Porphyridium* sp has the ability to reduce heavy metals pollution in the environment. This research was conducted in order to find out the potential use of *Porphyridium* sp for phycobiliproteins production and its ability in reducing the heavy metals Fe^{3+} and Co^{2+} in culture media. In this research, *Porphyridium* sp was cultivated in F2 non-silicate medium. The growth curve of *Porphyridium* sp was analyzed by cultivation at room temperature for 7 days and the OD 690 was monitored every day. Phycobiliproteins production was evaluated by analyzing the absorbance of the extracts at the wavelengths of 565, 620, and 650 nm. The ability of *Porphyridium* sp in the reduction of heavy metals Fe^{3+} and Co^{2+} was analyzed by AAS and calculated as bioconcentration factor (BCF). The data showed that the growth of *Porphyridium* sp in the media containing heavy metals Fe^{3+} and Co^{2+} was faster than in the control. The highest content of phycoerythrins, phycocyanins, and allophycocyanins of *Porphyridium* sp was reached after 4 days of cultivation with the addition of Fe^{3+} at 6.30 ppm. *Porphyridium* sp reduced Fe^{3+} and Co^{2+} in media with BCF values of 10.5804 and 0.1151, respectively.

1. Introduction

Indonesia is rich in aquatic biological resource such as microalgae or phytoplankton. Phytoplankton growth is rapid and its productivity is high. Phytoplankton or microalgae produce biomass up to 50 times than other higher plants. In addition, microalgae can live in different types of environment even in areas with limited land. Therefore, it does not compete with other plants [1]. *Porphyridium* sp, the most abundant species of red microalga of Rhodophyta division, has been the subject of intensive study by our group for a number of years. The cells of this red microalga were encapsulated within a cell-wall polysaccharide complex [2]. This strain is unique in building up fluorescent phycobiliproteins, exopolysaccharides, long-chain polyunsaturated fatty acids, carotenoids (zeaxanthin, tocopherol, etc and vitamins during its metabolic processes. *Porphyridium purpureum* growth process is clearly dependent on culture medium composition. The usual artificial sea water (ASW) culture medium has been proven to be a very convenient support in many algae species growth and production [3].

Red marine microalga *Porphyridium* sp is a Rhodophyta, a source of valuable phycobiliproteins which assemble into an organized cellular structure, the phycobilisome. They absorb light over a wide range of wavelengths in the visible part of the spectrum, and transfer the excitation energy by less radiation processes to the reaction in the photosynthetic membranes for conversion to chemical energy. The phycobiliproteins can be divided into three main classes depending on absorption properties: phycoerythrins (PE) max 540–570 nm, phycocyanins (PC) max 610–620 nm, and allophycocyanins (APC) max 650–655 nm. PE is a valuable candidate in the design and characterization of light-sensing elements in biosensors due to its properties that can absorb the light in the range of 540-570 nm. Another interesting application of the phycobiliproteins is their use as natural dyes in foods and cosmetics replacing the synthetic dyes, since the latter are in general toxic, carcinogenic, and therefore unsafe [4]. R- Phycoerythrin has been analyzed for its various physico-chemical properties and the analyses revealed that R- Phycoerythrin from *Kappaphycus alvarezii* has characteristic affinity towards different metal ions, inhibitors, organic solvents, and preservatives at different monochromatic irradiances [5]. The characteristic affinity towards metal ions of microalgae is potential for environmental remediation. Heavy metals are required by microalgae for enzymatic process as trace elements (≥ 3 g/cm^3), but they are toxic in high concentration. Remediation by microalgae has advantages since these microalgae can be used as fertilizer after remediation process or as biofuels with low cost, simple and flexible in application, and with low maintenance. *Porphyridium* has a potential to be used in heavy metals bioremediation. *Porphyridium cruentum* is promising for bioremediation of heavy metals to overcome environmental pollution [6]. The potential of *Porphyridium* sp needs to be explored so it can be used for decreasing the heavy metals content in waste water treatment. The objective of this study was to evaluate the phycobiliproteins production of *Porphyridium* sp and its ability in reducing heavy metals Fe^{3+} and Co^{2+}.

2. Experimental

2.1. Materials

Porphyridium sp isolate was obtained from Microbiology Laboratory – Research Center for Oceanography, Indonesian Institute of Sciences. The modification of F2 medium was sea water enriched with the following additives (per L) : $NaNO_3$ 0.075 g, Na_2HPO_4 0.05 g, Na-EDTA 0.00436 g, $ZnSO_4.7H_2O$ 0.022 g, $Na_2MoO_4.H_2O$ 0.0063 g, $CuSO_4.5H_2O$ 0.0098 g, $MnCl_2.4H_2O$ 0.18 g. The medium was sterilized (120°C, 20 min) prior to use. Cyanocobalamin 1 mg, thiamin HCl 0.2 mg, and biotin 1 mg (per L) were added to the sterile medium. The concentration of $FeCl_3.6\ H_2O$ and $CoSO_4$ were of 3.15 ppm and 6.30 ppm, respectively.

2.2. Methods

2.2.1. Porphyridium *sp cultivation.* The cultivation of *Porphyridium* sp was carried out by modified method [7]. The steps of cultivation were as follow: (i) 10 ml stock culture was cultivated in a shaker incubator at 30°C for 7 days. (ii) The culture was then put into 250 ml of medium. *Porphyridium* sp was cultivated in bottles which were connected to an aeration pump (0.037 MPa) and were exposed to sunlight at room temperature. The experiments were carried out for 7 days. The absorbance of the culture (λ 690) was monitored every 24 h.

2.2.2. Extraction and estimation of phycobiliproteins. Porphyridium sp culture was extracted and analyzed by the modified method [5]. The analysis of phycobiliproteins content was carried out at the day 4 of culture. The culture was frozen for 3 min and heated to 45°C for 3 min and then homogenized for 1 min. This treatment was carried out four times before being centrifuged at 7000 rpm for 10 min. The amounts of PC, APC, and PE in extracts were calculated based on measurements of the absorbance at 565, 620 and 650 nm using the following equations [5].

$$\text{PC (mg/ml)} = \frac{(OD_{620\ nm} - 0{,}7OD_{650\ nm})}{7{,}38}$$

$$\text{APC (mg/ml)} = \frac{(OD_{650\ nm} - 0{,}19OD_{620\ nm})}{5{,}65}$$

$$\text{PE (mg/ml)} = \frac{(OD_{565\ nm} - 2{,}8\ [PC] - 1{,}34\ [APC]}{12{,}7}$$

2.2.3. *Determination of Fe^{3+} and Co^{2+} accumulation.* The analysis of Fe^{3+} and Co^{2+} in culture media and biomass of *Porphyridium* sp was carried out according to Soeprobowati and Hariyati with modifications [6]. Before analysis, the samples were destructed by 3 ml of concentrated HNO_3 and heated at 50°C until there was no brown smoke. Samples were then cooled, added with 1.5 ml of H_2O_2, heated again until the volume was about ± 1 ml, and transferred to a 25 ml calibration flask. Analyses of Fe^{3+} and Co^{2+} were performed using a Shimadzu AA-7000 Atomic Absorption Spectrometer (AAS) with deuterium background subtraction. The wavelengths used for the analysis of Fe^{3+} and Co^{2+} were 248.16 nm and 240.41 nm, respectively.

The reduction of heavy metals was calculated, as well as *Porphyridium* sp population. Bioconcentration Factor (BCF) was calculated to determine the accumulation of heavy metals in the *Porphyridium* sp. BCF is a comparison between chemical concentrations in the organism and concentration in the environment or media [8].

$$BCF = C_{org} / C_{media}$$

C_{org} : heavy metals concentration in *Porphyridium* sp
C_{media} : heavy metals concentration in the culture media (after cultivation)

3. Results and Discussion

Porphyridium sp was cultivated for 7 days by using F2 non silicate medium and its growth was monitored every day. Data presented in Figure 1 and Figure 2 show that the growth of *Porphyridium* sp increased between day 2 and 3 in all curves. The growth of *Porphyridium* sp continued and the maximum of OD was at day 4. Then on day 5 a decline of microalgae populations *Porphyridium* sp started and continued until day 7.

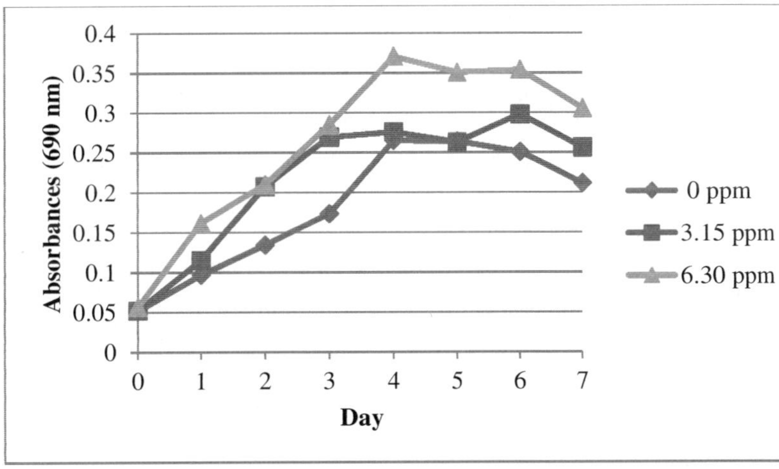

Figure 1. The effect of Fe^{3+} on the growth of *Porphyridium* sp.

Figure 1 shows that the concentration of Fe^{3+} influenced the growth of *Porphyridium* sp. The maximum OD (690 nm) was reached after 4 days of cultivation. This effect was also occurred when the medium contained Co^{2+}. *P. cruentum* tolerated a high concentration of heavy metal, concentration of 1 mg/L Cu had induced *P. cruentum* population growth. The life cycle of *Porphyrydium* tended to go a day forward in the higher heavy metal concentrations, but with a lower population [6].

The red microalga *Porphyridium* sp produces several high value polymers like polysaccharides, polyunsaturated fatty acids and phycobiliproteins which act as photosynthetic accessory pigments [9]. In this study, the phycobiliproteins content of *Porphyridium* sp was analyzed as well as its correlation with the heavy metals Fe^{3+} and Co^{2+} addition.

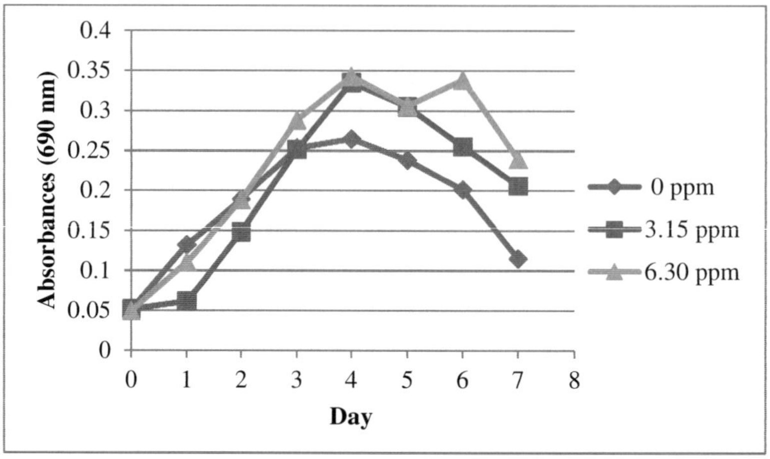

Figure 2. The effect of Co^{2+} on the growth of *Porphyridium* sp.

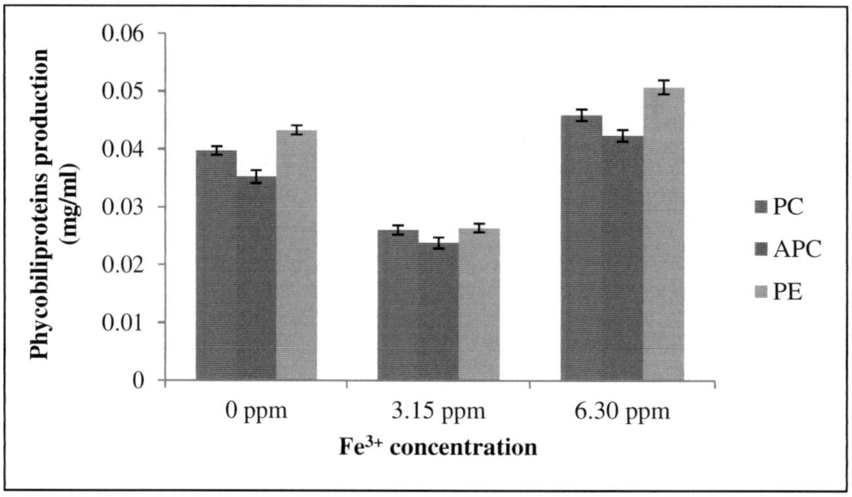

Figure 3. The effect of Fe^{3+} on phycobiliproteins production of *Porphyridium* sp (4 days cultivation, room temperature).

Figure 3 shows that with the addition of 3.15 ppm of $FeCl_3$ to the *Porphyridium sp* culture, the phycobiliproteins production was reduced. However, when the concentration of $FeCl_3$ was increased until 6.30 ppm, phycoerythrin (PE) content increased to 17.25%. This perhaps because the biomass was not homogeneous due to the aeration system in the cultivation process did not work well. Pugalendren *et al* [5] studied the addition of 0.1-10 mM $FeSO_4$ to the macroalgae *Kappaphycus alvarezii*, which resulted in the activity of R-PE (R-phycoerythirin) was decreased.

The influence of $CoSO_4$ addition to the *Porphyridium* sp culture was also investigated in regard to phycobiliproteins production. Figure 4 shows that the addition of $CoSO_4$ to *Porphyridium* sp increased phycobiliproteins production. PE production increased to almost ten times. The effect of cobalt on growth, pigment, and photosynthesis in unicellular green alga *Monoraphidium minutum* and *Nitzschia perminuta* was studied by El Sheekh *et al* [10]. Growth and pigment content were slightly increased at low concentration. The application of 0.5 and 1.5 ppm Co^{2+} increased *chlorophyll a* content by 4 and 12%, respectively. In this study, Co^{2+} at 3.15 ppm increased phycocyanins (PC) and phycoerythrin (PE) contents. Co^{2+} is a trace element that is important for pigment production such as *chlorophyll a* and phycoerythrin.

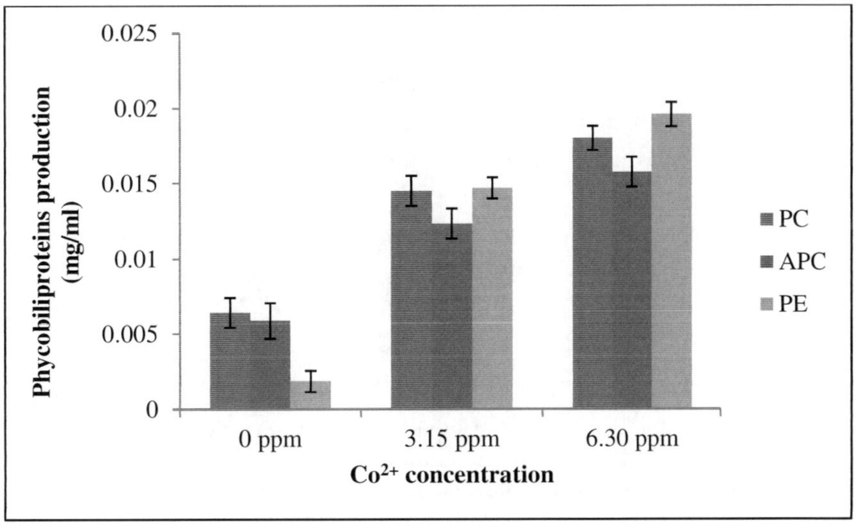

Figure 4. The effect of Co^{2+} on phycobiliproteins production of *Porphyridium* sp (4 days cultivation, room temperature).

The use of microalgae for bioremediation has been studied by many researchers. *Porphyridium* had a potential to be used in heavy metals bioremediation. *P. cruentum* was qualified for bioremediation due to the absence of toxic production, was easily cultured, had the ability to grow in extremes of salinity, pH, and temperature, rapid growth in defined media, ability to achieve a high population, and was easily harvested [11]. Bioaccumulation of Fe^{3+} and Co^{2+} by *Porphyridium* sp has been studied and the data was presented on Table 1.

Table 1. Fe^{3+} and Co^{2+} content (ppm) in media and biomass of *Porphyridium* sp (4 days cultivation, room temperature).

Concentration of heavy metals in media (ppm)	Fe^{3+} (ppm), after cultivation		Co^{2+} (ppm), after cultivation	
	Media	Biomass	Media	Biomass
$FeCl_3$, 0	0.5421	0.4404		
$FeCl_3$, 6.30	0.6203	6.5630		
$CoSO_4$, 0			-0.0061	-0.0147
$CoSO_4$, 6.30			3.8700	0.4454

The data in Table 1 shows that for the control culture (Co:0 ppm), the Co^{2+} contents in the medium and biomass were negative. Meanwhile, Fe^{3+} contents in the medium and biomass (Fe:0 ppm) were positive. We predicted that *Porphyridium* sp culture contained Fe^{3+}. By the addition of 6.30 ppm of $FeCl_3$, Fe^{3+} content in biomass was much higher than in medium. After added by 6.30 ppm of $CoSO_4$, Co^{2+} content in biomass was lower than in medium. This assumed that *Porphyridium* sp culture could absorb Fe^{3+} optimally. In order to assess the accumulation properties of the used *Porphyridium* sp, bioconcentration factor (BCF) was determined, indicating the ratio of heavy metal concentration in biomass versus heavy metal concentration in medium. BCF values of Fe^{3+} and Co^{2+} in *Porphyridium* sp are represented in Figure 5.

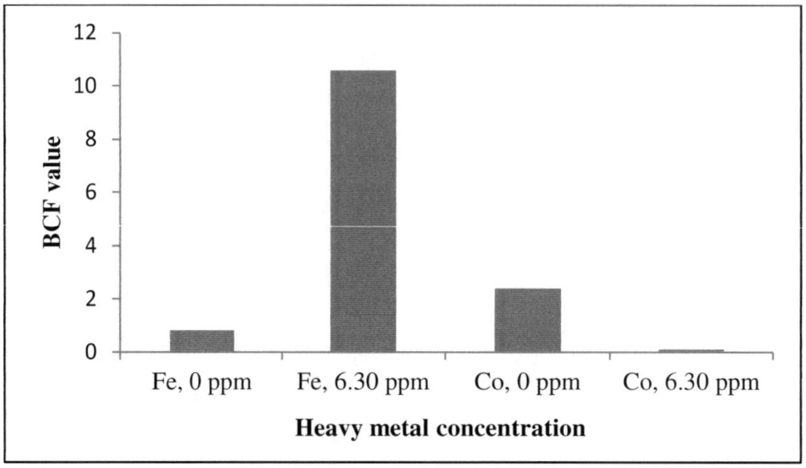

Figure 5. BCF (bioconcentration factor) value of Fe^{3+} and Co^{2+} in *Porphyridium* sp (4 days cultivation, room temperature).

The data show that BCF value of culture sample which contained 6.30 ppm Fe^{3+} (10.58) was much higher than the culture contained Co^{2+} (0.1151). It was shown from this study that *Porphyridium* sp successfully demonstrated the absorption and removal of Fe^{3+} from the medium. This was related to the heavy metal release rate was relatively lower than its absorption. Heavy metals absorption occurred in 2 ways i.e. heavy metal ionic change with cell wall caption or development of covalent bound between heavy metals with active ionic cell wall. *P. cruentum* cell wall consisted of organic protein, polysaccharide, alginate acid and urinate acid which were able to bind heavy metals [12].

There was a correlation between phycobiliproteins production and the accumulation of heavy metals (BCF) by *Porphyridium* sp; if phycobiliproteins produced were low, the BCF also had a low value. Phycoerythrin content of the culture which contained 6.30 ppm Fe^{3+} was higher compared to the culture contained Co^{2+} and this also occurred to BCF value of the same culture sample. This assumed that Fe^{3+} stimulated the photosynthetic process of microalgae. BCF was calculated as the homeostatic ratio of heavy metal concentration on the *P. cruentum* to heavy metal concentration on the media. Microalgae had a protection mechanism against heavy metals by development of heavy metals complex with cellular protein without changing its activity [12].

4. Conclusions

The growth of *Porphyridium* sp was studied and evaluated, as well as its phycobiliprotein production and heavy metals Fe^{3+} and Co^{2+} reduction ability. The addition of Fe^{3+} (6.30 ppm) and Co^{2+} (6.30 ppm) increased the phycobiliproteins (PC, APC, PE) production. BCF value of culture sample which contained 6.30 ppm Fe^{3+} (10.58) was much higher than the culture contained Co^{2+} (0.1151). *Porphyridium* sp successfully demonstrated the absorption and removal of Fe^{3+}.

Acknowledgements

This research was supported by Research Unit for Clean Technology and Research Centre for Geotechnology - Indonesian Institute of Sciences (LIPI).

References

[1] Kwangdinata R, Raya I and Zakir M 2014 *ISRN Renew. Energy* **2014** ID 107278
[2] Geresh S, Arad S M, Levy-Ontman O, Zhang W, Tekoah Y and Glaser R 2009 *Carbohydrate Res.* **344** 343–9
[3] Velea S, Ilie L and Filipescu L 2011 *UPB Sci. Bull. B* **73** 81-94

IOP Conf. Series: Earth and Environmental Science **160** (2018) 012006 doi:10.1088/1755-1315/160/1/012006

[4] Roman R B, Alvarez-Pez J M, Fernandez F G A and Molina-Grima E 2002 *J. Biotechnol.* **93** 73–85
[5] Pugalendren S, Sarangam B and Rengasamy R 2012 *J. Acad. Indus. Res.* **1** 407-11
[6] Soeprobowati T R and Hariyati R 2013 *Int. J. Marine Sci.* **3** 212-8
[7] Priatni S, Kosasih W, Budiwati T A and Ratnaningrum D 2017 *IOP Conf. Series: Earth Environ. Sci.* **60** 012009
[8] Ivanciuc T, Ivanciuc O and Klein D J 2006 *Molec. Divers.* **10** 133-45
[9] Csogör Z, Kiessling B, Perner I, Fleck P and Posten C 2001 *J. Appl. Phycology* **13** 317–24
[10] El Sheekh M M, El Nagar A H, Osman M E H and Mazali E 2003 *Braz. J. Plant Physiol.* **15** 159-66
[11] Wilde E W, Radway J C, Domingo J S, Zingmark R G and Whitaker M J 1988 *Final Report for TTP SR-16-PL-42 (Formerly SR-141019) Bioremediation of Aqueous Pollutants Using Biomass Embedded in Hydrophilic Foam (U)* US Department of Energy
[12] Wang J and Chen C 2009 *Biotechnol. Adv.* **27** 195-226

IOP Conf. Series: Earth and Environmental Science **160** (2018) 012007 doi:10.1088/1755-1315/160/1/012007

Scaling up process for fish peptone production

W Kosasih*, D Ratnaningrum, E S Endah, S Pudjiraharti and S Priatni

Research Unit for Clean Technology, Indonesian Institute of Sciences (LIPI) Komplek LIPI Bandung, Jalan Sangkuriang, Gedung 50, Bandung 40135, Indonesia

*Email: wawankosasih@yahoo.co.id

Abstract. Fish peptone is one of protein hydrolysate products that can be used as nitrogen source for microbial growth. Peptone can be made from marine fish whereas protease enzyme was made from either papaya extract or commercial papain. Fish peptone is not only an eco-friendly approach but also can produce economic materials used in formulation of culture media. The objective of this research was to optimise the scaling up process of fish peptone production. An amount of 200 g of clean fish marine was blended with 1 L of water and added with 0.2 % of commercial papain at pH 8. Peptone production was carried out in a 1 L glass reactor with agitation (50, 100, 150, and 200 rpm) for 1-8 h. Fish slurry was then incubated at 50°C for 8 h. Fish protein hydrolysate was spray-dried with inlet and outlet temperature of 90°C and 80°C, respectively, at flow rate of 200 mL/h. Results revealed that the best condition for hydrolysis was shown by protein soluble content of 17.1 mg/L, α-amino content of 147 mM, and degree of hydrolysis of 8.3% which was achieved by agitation at 100 rpm for 8 h reaction. The yield of dried peptone product was 11.9 % (w/v). This peptone powder had higher soluble protein and α-amino content (442.3 mg/g, 124 mM) than the commercial peptone (258.5 mg/g, 65 mM). The microbial growth test of fish peptone products showed similar effectiveness compared with commercial bacterial peptone for the growth of *Escherichia coli* and *Streptococcus aureus*.

1. Introduction

Peptone is one of the ingredients used in the manufacture of microbial growth media. Peptone provides nitrogen for bacteria to grow. Peptone as an intermediate material between peptides and proteins is soluble in water, not coagulated by heat, and can be precipitated by ammonium sulphate and zinc sulphate [1]. Peptone can be extracted from animal proteins in the form of internal organs of meat, gelatine, milk, casein, plants, and khamir [2,3]. Due to its increasing demand and local peptone industry has not yet been developed, Indonesia imported 5.102 tonnes of peptone in 2013 at the value of US$ 20.76 million. The import increased compared to the year 2012, which was 3.296 tonnes at the value of US $ 12.15 million [4]. Fish peptone has an important economic value compared with other side products such as fish silage and fish meal [5].

Fish protein hydrolysates (peptone) were made by hydrolysing fish using proteolytic enzymes such as papain, ficin, trypsin, pancreatin, pronase, or enzymes isolated from proteolytic microorganisms [6]. Enzymatic processes do not require high temperature, it hydrolyses polypeptide specifically, and conserves all existing amino acids. This provides an advantage compared with acid hydrolysis that takes place at high temperatures, consequently damages either some or all amino acids. In addition, acid hydrolysis results in peptones with high salt content due to the formation of salts during the neutralization process [6].

Peptone production using enzyme from papaya latex has been carried out previously [7]. However, its practical application was difficult due to enzyme resource derived from dried papaya latex. Moreover, application of dried papaya latex based-enzyme in a large/industrial scale is ineffective since it is costly and impractical. Commercial enzymes are easier and cheaper compared to dried papaya latex with better quality and supply. The use of low valued fish for peptone production is promising because it is less expensive as well as provide better quality compared with commercial peptone.

This research aimed to produce peptone in a 1 L apparatus using commercial papain enzyme, commonly used in pilot scale.

2. Method

2.1. Enzyme preparation

A commercial enzyme (brand Xian Arisun ChemParm Co. Ltd, CAS no. 2323.627-2, Shaanxi, China, in the form of powder) was used in the experiment. A stock solution was made by dissolving the enzyme with aquadest at the concentration of 10% (w/v).

2.2. Enzymatic reaction

To investigate the influence of enzyme concentration, enzymatic reaction was performed with commercial papain enzyme at the concentration of 0.1, 0.2, 0.3, 0.4, and 0.5 %. The constant concentration of 25% substrate was used and the total volume was 40 mL.

To investigate the influence of substrate concentration, enzymatic reaction was performed with substrate concentration of 5, 10, 15, 20, and 25 %. Commercial papain enzyme at constant concentration of 0.2% was used and the total volume was 40 mL.

2.3. Peptone powder production

An amount of 400 g of frozen fish was thawed, washed, and blended with 1.5 L of water. The fish slurry was then adjusted to pH 8 by adding 2 M NaOH. Aquadest was added to reach total volume of 1,960 mL. Fish porridge was put into a reactor and 20 mL of stock enzyme solution (10% concentration) was added, thus the final enzyme concentration was 0.2%. The mixture was heated at 50°C and stirred at 50, 100, 150, or 200 rpm for 8 h each. Enzymatic reaction was stopped by heating samples in boiling water for 10 min.

Protein hydrolysate was separated from its bone and fish scales by vacuum filtration using a Büchner flask coated with filter paper. The clear protein hydrolysate was spray-dried with inlet and outlet temperature of 90 and 80 °C, respectively and a flow rate of 200 mL/h.

2.4. Analysis of α-amino content

Analysis of α-amino content was performed using formol titration method [8]. Into a 100 mL beaker glass were added 5 mL of sample, 50 mL of aquadest, saturated K-oxalate solution in water (1: 3), and one drop of 0.01 % rosalin-chloride indicator. The mixture was adjusted to pH 8.2 by adding 0.1 M NaOH dropwise while stirred. A 10 mL of formalin (35%, v/v) was added. The titration was continued with the addition of 0.1 M NaOH to pH 9.2. The NaOH volume added was recorded as the amount of titre. Calculation of α-amino content is as follows:

$$\%\,\alpha - amino\ (\%N) = \left(\frac{\Delta V}{sample\ weight \times 10}\right) N.NaOH \times 14.008$$

Where:

%N = Amino concentration
ΔV = Added NaOH volume (titer)
N. NaOH = NaOH concentration (mol/L)

The degree of hydrolysis (% DH) was calculated as follows:

$$\%DH = \frac{\alpha - Amino}{Total\ N} \times 100\%$$

Total N was determined by the Kjeldahl method [9].

2.5. Analysis of soluble protein content
The analysis of the total protein of fish peptone was carried out using Lowry's method [10]. Lowry A solution consisted of 4 mg/mL of NaOH and 20 mg/mL of Na_2CO_3 in water. Lowry B solution consisted of 10 mg/mL of potassium sodium tartrate and 5 mg/mL of $CuSO_4$ in water. Lowry C solution was prepared by mixing 50 mL of Lowry A solution and 1 mL of Lowry B solution. Lowry D solution was prepared by diluting Folin-Ciocalteu reagent with aquadest at equal volumes. As much as 500 µl of fish peptone samples were added with 2.5 mL of Lowry C solution and dissolved for 10 min. Samples were then added with 0.5 mL of solution D and dissolved for 30 min. Sample solution was measured with a spectrophotometer at 500 nm. A series concentration of bovine serum albumin (BSA) was used for standard curve of protein.

3. Results and discussion
The results of peptone production in a 1 L scale using commercial enzyme Xian Arisun ChemPharm are as follows:

3.1. Effect of enzyme and substrate concentration
The effects of varying enzyme and substrate concentration was observed as the formation of α-amino. The results are shown in Table 1 and 2.

Table 1. Effect of enzyme concentration on α-amino concentration.

Enzyme concentration (%)	α-amino concentration (%N)
0.1	0.516 ± 0.03
0.2	0.521 ± 0.02
0.3	0.544 ± 0.02
0.4	0.555 ± 0.02
0.5	0.527 ± 0.02

Table 2. Effect of substrate concentration on α-amino concentration.

Substrate concentration (%)	α-amino concentration (mM)
5	36.7 ± 3.05
10	51.3 ± 4.62
15	79.3 ± 20.13
20	102.7 ± 16.17
25	126.7 ± 32.33

Table 1 shows that varying enzyme concentration did not have significant effect on the resulting α-amino concentration. On the other hand, every 5% increase of substrate concentration resulted in significant increase of α-amino concentration (Table 2). The highest increase was observed at 20% substrate concentration.

Based on results shown in Table 1 and 2, 0.2% enzyme concentration and 20% substrate concentration were used in peptone production. pH 8 was selected as the optimum pH from a preliminary experiment and a previous research using the same substrate and papain enzyme from papaya latex [7].

3.2. Effect of time and agitation on fish hydrolysate production

The effect of time and agitation on fish hydrolysate production was observed as the formation of α-amino. The results are shown in Figure 1.

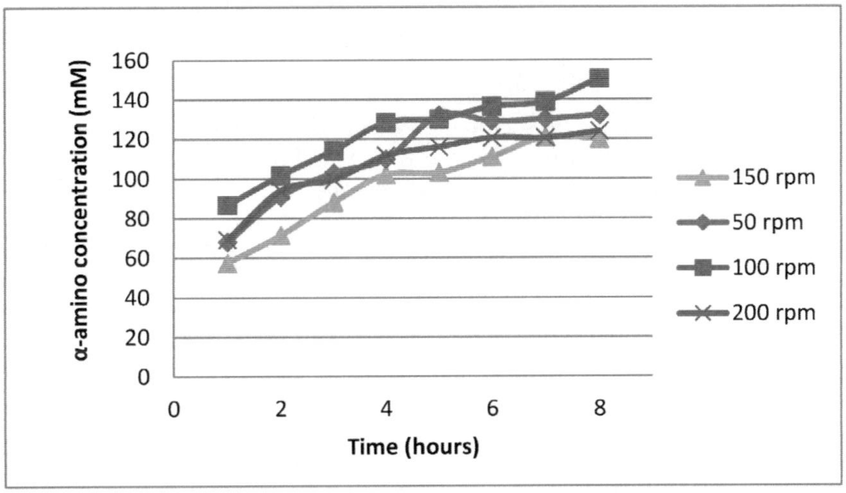

Figure 1. The effect of agitation and enzymatic reaction time on α-amino concentration of fish peptone.

During the 8 h reaction time, agitation of 100 rpm generally produced higher α-amino concentrations compared with other agitation speeds. The highest concentration of α-amino from 100 rpm agitation was obtained at 8 h with concentration of 150 mM. Agitation that produced the lowest α-amino was 150 rpm (123 mM at 8 h). Concentrations of α-amino from agitations of 50 rpm and 200 rpm were relatively similar up to 4 h, while the maximum α-amino concentrations at 8 h were 138 mM (50 rpm) and 124 mM (200 rpm).

3.3. Peptone powder production by spray dryer

Fish peptone resulted by the agitation of 100 rpm was spray-dried with inlet and outlet temperature of 90 and 80 °C, respectively, and flow rate of 200 mL/h. This condition was a modification from Lee [11]. The yield was 20.4 g and its powder performance is shown in Figure 2.

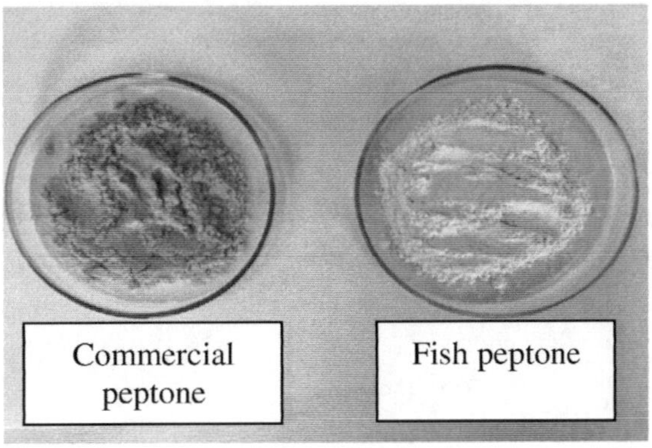

Figure 2. Fish peptone and commercial peptone.

IOP Conf. Series: Earth and Environmental Science **160** (2018) 012007 doi:10.1088/1755-1315/160/1/012007

To determine the protein content of fish and commercial peptones, as much as 1 g of each peptone was dissolved in 25 mL of aquadest (ratio1:25). Protein and α-amino contents of the peptone solutions are presented in Table 3.

Table 3. Protein and α-amino content.

Sample (1:25)	Protein content (mg/mL)	α-amino content (mM)
Fish peptone	17.69	124
Commercial peptone	10.34	65

The protein contents were calculated based on the data in Table 1. The protein content in fish peptone powder was 442.25 mg/g whereas the commercial peptone was 258.50 mg/g. Thus, protein and α-amino contents of fish peptone were higher than commercial peptone.

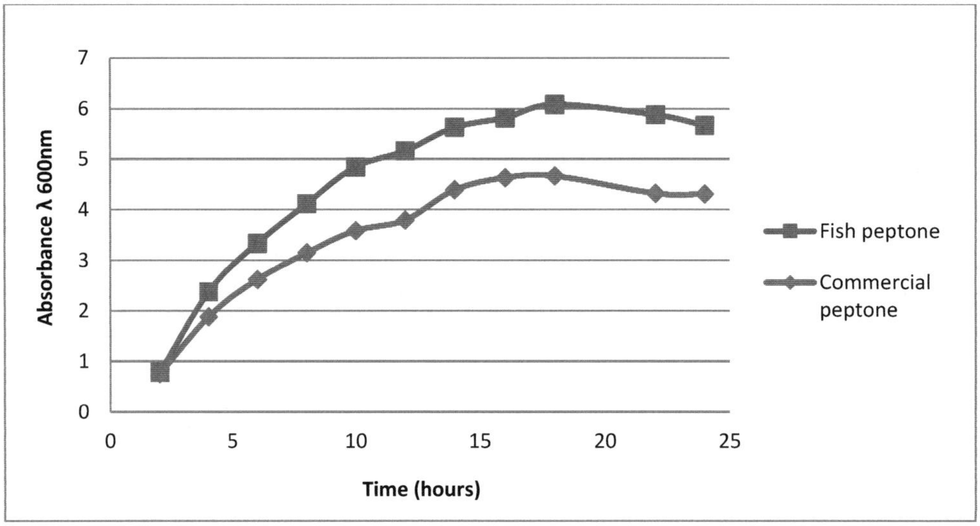

Figure 3. The growth curves of *E. coli* in microbial growth media.

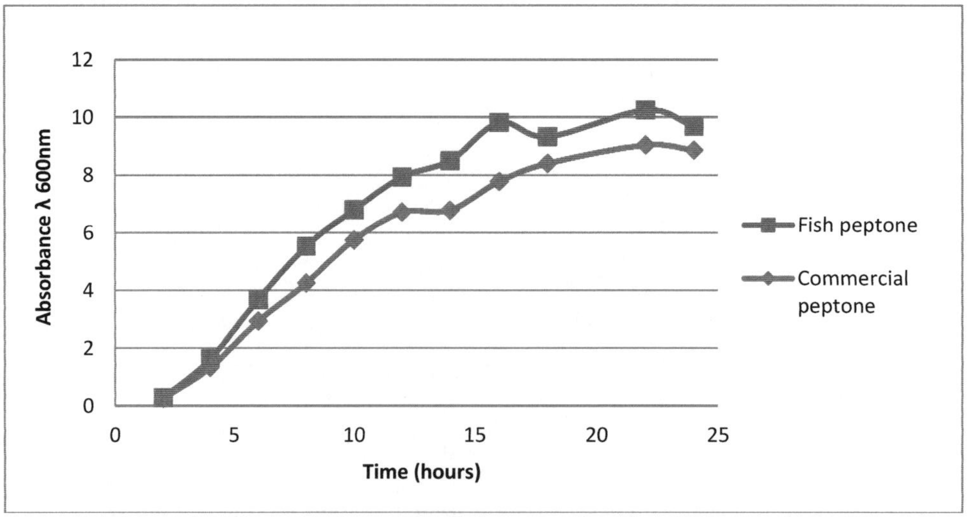

Figure 4. The growth curves of *S. aureus* in microbial growth media.

3.4. Microbial growth

The suitability of peptone fish as a source of nitrogen for microbial growth media was tested using *E. coli* and *S. aureus*. The microbial growth media consisted of 2% of fish or commercial peptone, 2% NaCl, and 1% yeast [12]. Measurement was carried out every two hours at 600 nm. Growth curves for *E. coli* and *S. aureus* can be seen in Figure 3 and Figure 4.

E. coli grown in media containing fish pepton showed a higher cell density compared with commercial peptone. The data shows that lag phase of *E. coli* in peptone was about 2 h. Log phase of *E.coli* in fish peptone and commercial peptone occurred for 18 h with maximum absorbances (A_{600}) of 6.084 and 4.668, respectively. After 18 h, the absorbance of *E. coli* decreased indicating the death phase had occurred. Vieira *et al* [13] reported the performance of peptones produced from different fisheries by-product for *E. coli* compared with commercial peptone. The better performance of fish peptone in comparison with commercial media was also reported by other researches [14].

As shown on Figure 4, the growth of *S. aureus* in fish peptone was higher than commercial peptone. The data shows that lag phase of *S. aureus* in peptone was about 4 h. Log phase of *S. aureus* in fish peptone and commercial peptone occurred for 22 h with maximum absorbances (A_{600}) of 10.248 and 9.036, respectively. The lag phase is defined as the transition to the exponential phase after the initial population has doubled. The lag phase is thought to be due to the physiological adaptation of the cell to the culture conditions. This may involve a time requirement for induction of specific messenger RNA (mRNA) and protein synthesis to meet new culture requirements. Meanwhile, during log phase or exponential growth the rate of increase of cells in the culture is proportional to the number of cells present at any particular time and the death phase is characterized by a net loss of culturable cells [15].

4. Conclusions

Low valued fish can be made into peptone using commercial papain enzymes. Peptone from fish has a high protein content for *E. coli* and *S. aureus* growth.

Acknowledgements

This research was supported by DIPA Project 2017 - Research Unit for Clean Technology, Indonesian Institute of Sciences (LIPI).

References

[1] Neidhardt F C, Bloch P L and Smith D F 1974 *J. Bacteriol.* **119** 736–47
[2] Wisuthiphaet N, Kongruang S and Chamcheun C 2015 *J. Med. Bioeng.* **4** 466–70
[3] Al Bahri M B A G, Al Naimi S A and Ahammed S H 2009 *J. Al-Khwarizmi Eng.* **5** 1-19
[4] Badan Pusat Statistik 2014 *Tabel Impor Menurut Komoditi*
[5] Ariyani F, Heruwati E S, Murdinah, Wibowo T B S and Susetyo A R 2001 *J. Penelitian Perikanan Indonesia* **7** 75-84
[6] Dufossë L, De La Broise D and Guerard F 2001 *Curr. Microbiol.* **42** 32-8
[7] S Priatni, W Kosasih, T A Budiwati and D Ratnaningrum 2017 *IOP Conf. Ser.: Earth Environ. Sci.* **60** 012009
[8] Sudarmadji S 1997 *Prosedur Analisa Bahan Makanan dan Pertanian* (Yogyakarta: Liberty)
[9] AOAC 1995 Official Methods of Analysis of the Association of Analytical Chemists (Washington DC: AOAC)
[10] Lowry O H, Rosebrough N J, Farr A L and Randall R J 1951 *J. Biol. Chem.* 193 265-75
[11] Lee G 2002 Rational Design of Stable Protein Formulations, ed Carpenter J F and Manning M C (New York: Springer) pp 135-58
[12] Becton, Dickinson and Company 2009 DifcoTM & BBLTM Manual of Microbiological Culture Media 2nd edition, ed Zimbro et al (Sparks: BD Diagnostics)
[13] Vieira, G H, Vieira R H, Macrae A and Sousa O V 2005 *J. Sci. Food Agric.* **85** 1235–7
[14] Fallah M, Bahram S and Javadian S R 2015 *Food Sci. Nutr.* **3** 153–7
[15] Maier R M 2009 *Environmental Microbiology* pp 38-54 (Academic Press. Inc.)

IOP Conf. Series: Earth and Environmental Science **160** (2018) 012008 doi:10.1088/1755-1315/160/1/012008

Utilization of soybean curd residue for carbon-based adsorbent material and its characterization

A K Fikriyyah[1], E R Chaldun[1*] and A Hardiansyah[2]

[1] Research Unit for Clean Technology, Indonesian Institute of Sciences (LIPI) Komplek LIPI Bandung, Jalan Sangkuriang, Gedung 50, Bandung 40135, Indonesia
[2] Institut Teknologi dan Sains Bandung
Jl. Ganesha Boulevard, Cikarang Pusat,Bekasi, Jawa Barat 17530

*Email: elsy002@lipi.go.id

Abstract. Soybean curd residue (SCR) as carbon sources is able for absorbance materials. The utilization of SCR has provided new opportunities and challenges for advanced materials research. The transformation of the carbon structure of the SCR provided graphene-like structures with high surface area and metal ion absorbing capability. The carbon-based adsorbent material was prepared by pyrolysis method. The transformation of carbon structure to produce graphene structure was carried out by Hummers method. Magnetization was carried out by the addition of $FeCl_3.6H_2O$ and $FeCl_2.4H_2O$ to form a magnetite compound (Fe_3O_4) and was applied as absorbance of metal ions. Separation process was carried using magnetic field. The samples were characterized by scanning electron microscopy (SEM), fourier-transform infrared (FTIR), Brauner Emmet Teller (BET) and atomic absorption spectrophotometry (AAS). The results showed that the carbon from the SCR had graphene like-structure which was able to absorb nickel metal ions.

1. Introduction

Fabrication of graphene from natural and industrial waste materials has been carried out by several researchers [1–3]. Various natural wastes have been applied to make graphene such as sugar, jute, rice husk [3], wheat straw [4], wood, leaf, fruit wastes, bagasse, animal wastes and industrial wastes [2]. It shows that graphene can generally be made from high carbon materials such as carbohydrates and proteins. Li *et al* reported that SCR contained carbohydrates in the form of cellulose at about 50 wt% [5].

Graphene is one of carbon structures with many advantages such as high surface area [6], good electrical and thermal conductivity [7], and excellent mechanical flexibility [8]. Graphene can be used as an adsorbent on metal waste. Magnetic material-decorated graphene namely magnetite reduced graphene oxide (MRGO) as adsorbent for metal ion waste has been previously synthesized [9–11]. MRGO is produced to develop the separation process for liquid waste by magnetic separation technique. The process used less energy and was environmentally friendly.

In this research, SCR was used to produce graphene-based structures for MRGO as adsorbent through magnetic separation technique. Scanning electron microscopy and energy dispersive X-ray spectroscopy (SEM-EDS) was used to characterize morphology and elemental analysis of MRGO. Fourier-transform infrared (FTIR) was used to analyze chemical structures and the functional groups of graphene and MRGO. Surface area and pore size of graphene were analyzed by Brauner Emmet

IOP Conf. Series: Earth and Environmental Science **160** (2018) 012008 doi:10.1088/1755-1315/160/1/012008

Teller (BET) whereas the quantity of metal ion adsorption was carried out by atomic absorption spectrophotometry (AAS).

2. Materials and Method

2.1. Materials and chemicals
Carbon from SCR was processed by pyrolysis at 700 °C for 1 h. This method was adopted from a previous method with minor modification [12]. Graphene was synthesized form carbon-based SCR by Hummers method as described in the previous procedure [2]. $FeCl_3.6H_2O$ and $FeCl_2.4H_2O$ were purchased from Sigma-Aldrich to build Fe_3O_4 particle on MRGO. Sulphuric acid, nitric acid, potassium permanganate, hydrochloride acid, hydrogen peroxide, ammonia and ethanol were purchased from Merck Company. Deionized water (DI water) was used for aqueous solutions and rinsing purpose.

2.2. Synthesis of MRGO
MRGO was synthesized by impregnation of Fe_3O_4 particles on graphene as described by previous method [13]. A quantity of 0.3 g of graphene was dissolved in 150 mL DI water and activated using ultrasonic bath for 30 min. Mixture of 25 mL of $FeCl_3.6H_2O$ and $FeCl_2.4H_2O$ solution (molar ratio 2:1) was added to graphene under stirring. A 25%wt ammonia solution was added until pH~10 and the temperature was increased up to 80–85 °C for 45 min. The mixture was filtered and washed with ethanol and deionized water several times until neutral condition was achieved. The suspension was oven-dried at 45 °C.

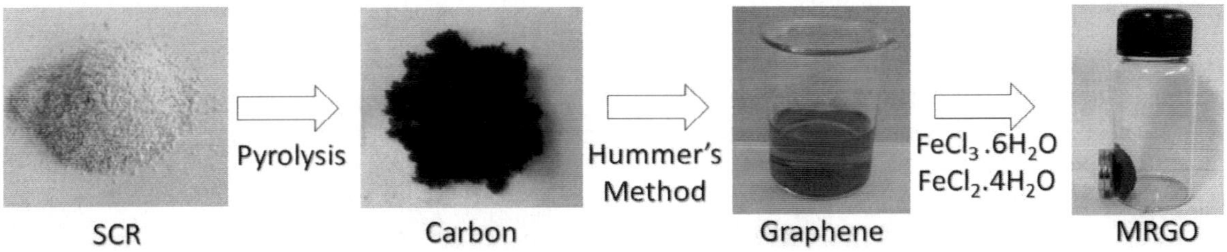

Figure 1. A schematic diagram of MRGO synthesis.

2.3. Application of MRGO in nickel ion waste
MRGO and nickel ion waste solution were mixed with ratio (w/v) of 1:2000. The mixture of MRGO-nickel was shaken for 6 h. Then, MRGO was separated by attaching a magnetic field for 30 min.

2.4. Characterization
The morphology image and elemental identification of the samples were obtained with SEM-EDS (JEOL JSM-IT300) under the operating voltage of 20 kV. The surface area of graphene was measured by BET (Quantachrome NOVA 4200e). FTIR spectra (Nicolet iS5) was recorded to know the functional groups and the existing bonds on MRGO. AAS (SHIMAZU AA6300) were performed to check the quantity of nickel ions.

3. Results and Discussion
Figure 2 illustrates the morphology and elemental analysis of MRGO using SEM-EDS. It shows that the matrix of Fe_3O_4 particle and wrinkle layers could be considered on MRGO surface. The appearance of iron (Fe) element is shown on EDS data. This indicated that Fe particles had been successfully distributed on the surface of MRGO. The EDS data also illustrates other elements such as carbon (C) and oxygen (O). The peak of O element was related to Fe particles from Fe_3O_4.

IOP Conf. Series: Earth and Environmental Science **160** (2018) 012008 doi:10.1088/1755-1315/160/1/012008

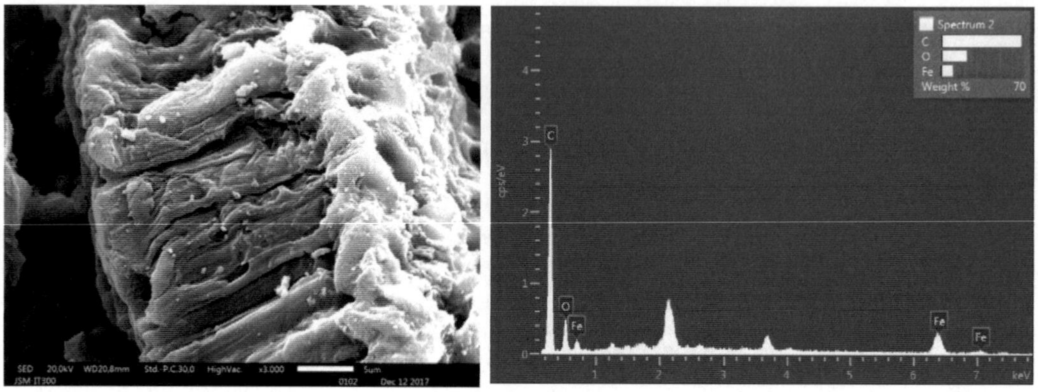

Figure 2. SEM images and EDS data of MRGO.

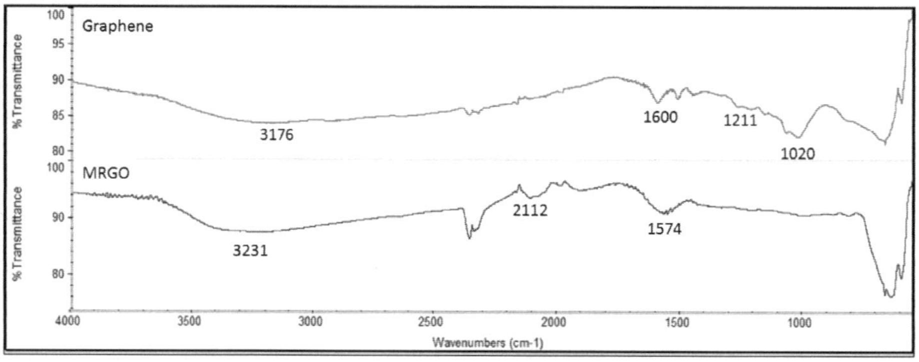

Figure 3. FTIR spectra of graphene and MRGO.

Figure 3 reveals FTIR spectra of graphene and MRGO. FTIR spectra of graphene shows the bands at 1020 cm^{-1} (C=O), 1211 cm^{-1} (C-O), 1600 cm^{-1} (C=C) and 3176 cm^{-1} (O-H). These bands indicated that graphene had been oxidized because the presence of carboxyl, epoxy group, and alkoxy [14]. Then, FTIR spectra of MRGO exhibited the band at 3231 cm^{-1}(O-H) and derivation at the intensity of the adsorption bands of oxygen-containing functional groups [15]. It means that graphene oxide had been reduced on MRGO. Besides, MRGO also showed new bands at 1574 cm^{-1} and 2112 cm^{-1} shows that COO-symmetric vibration was connected with Fe^{3+} [16] and skeletal vibration of graphene sheet [17].

Table 1. BET measurement of graphene.

Surface Area (m^2/g)	Pore Size (nm)	Total Pore Volume (cc/g)
151.472	1.178	0.0892

To determine the surface area and pore size of the graphene, BET measurements were performed. Table 1 shows BET measurements of graphene. Graphene in MRGO had a surface area of 151.472 m^2/g. Based on a reference, graphene oxide had a surface area of 318 m^2/g, represents a decrease in the surface area of the graphene that agglomeration occurred when MRGO was synthesized [11]. In this case, graphene made from carbon-based SCR had a surface area that was not much different with graphene from commercial graphite. Therefore, carbon-based SCR can be an alternative to produce graphene.

To test the absorbance ability, MRGO was applied in nickel metal ion waste. The absorbance result was measured using AAS to determine the alteration of Ni metal ion content in the waste (Table 2).

Table 2. AAS measurements of nickel ion content in waste solution MRGO.

Treatment	Absorbance (a.u)	Ni Content		Ni Adsorbed	
		(ppm)	(%)	(ppm)	(%)
Before	0.0304	77188	100	0	0
After	0.027	67572	87.54	9616	14.23

The result exhibits that there was a reduction of Ni ions before and after treatment. A total of 14% of Ni ion content was absorbed by MRGO. The metal ion adsorption mechanism occurred due to the covalent bonding of the carboxyl group (-COOH) on the graphene surface. The carboxyl group had a negative charge that could bind to nickel ions (Ni^{2+}) [18]. When an external magnetic field was applied, there was an interaction between Ni^{2+} and MRGO. The interaction process of Ni^{2+} that bound to carboxyl group on the graphene surface was subjected to a pull driven by magnetic particles. This caused a decrease of Ni ions content in the solution because it was adsorbed on MRGO. It shows that graphene from SCR can be used as an adsorbent for metal ion waste which have toxic properties. Inexpensive and environmentally friendly materials can be utilized as potential adsorbents for treating water sources contaminated by heavy metals.

4. Conclusion

Utilization of SCR to produce graphene-based structures for MRGO as adsorbent through magnetic separation technique was successfully developed. The metal ions can be decreased as much as 14% in aqueous solution after the application of MRGO and external magnetic field simultaneously.

References

[1] Somanathan T, Prasad K, Ostrikov K, Saravanan A and Krishna V 2015 *Nanomaterials* **5** 826–834
[2] Akhavan O, Bijanzad K and Mirsepah A 2014 *RSC Adv.* **4** 20441-20448
[3] Ojha K, Kumar B and Ganguli A K 2017 *J. Chem. Sci.* **129** 397–404
[4] Chen F, Yang J, Bai T, Long B and Zhou X 2016 *J. Electroanal. Chem.* **768** 18–26
[5] Li S, Zhu D, Li K, Yang Y, Lei Z and Zhang Z 2013 *ISRN Ind. Eng.* **2013** ID 423590
[6] Chae H K, Siberio-Pérez D Y, Kim J, Go Y, Eddaoudi M, Matzger A J, O'Keeffe M and Yaghi O M 2004 *Nature* **427** 523–527
[7] Balandin A A, Ghosh S, Bao W, Calizo I, Teweldebrhan D, Miao F and Lau C N 2008 *Nano Lett.* **8** 902–907
[8] Lee C, Wei X, Kysar J W and Hone J 2008 *Science* **321** 385–388
[9] Liu M, Chen C, Hu J, Wu X and Wang X 2011 *J. Phys. Chem. C* **115** 25234–25240
[10] Chandra V, Park J, Chun Y, Lee J W, Hwang I and Kim K S 2010 *ACS Nano* **4** 3979–3986
[11] Hoan N T V, Thu N T A, Duc H V, Cuong N D, Khieu D Q and Vo V 2016 *J. Chem.* **2016** ID 2418172
[12] Ma G, Ran F, Peng H, Sun K and Zhang Z 2015 *RSC Adv.* **5** 83129–83138
[13] Liu M, Chen C, Hu J, Wu X and Wang X 2011 *J. Phys. Chem. C* **115** 25234–25240
[14] Chandra V, Park J, Chun Y, Lee J W, Hwang I C and Kim K S 2010 *ACS Nano* **4** 3979–3986
[15] Chang Y P, Ren C L, Qu J C and Chen X G 2012 *Appl. Surf. Sci.* **261** 504–509
[16] Ge J, Hu Y, Biasini M, Beyermann W P and Yin Y 2007 *Angew. Chemie - Int. Ed.* **46** 4342–4345
[17] Qi T, Huang C, Yan S, Li X J and Pan S Y 2015 *Talanta* **144** 1116-1124
[18] Idzni A F 2016 *Sintesis Magnetite (Fe_3O_4) Reduced Graphene Oxide (MRGO) dari Grafit untuk Aplikasi Adsorben Ion Logam Limbah Elektroplating* (Bandung, Indonesia: Bandung Institute of Science and Technology)

IOP Conf. Series: Earth and Environmental Science **160** (2018) 012009 doi:10.1088/1755-1315/160/1/012009

The production of Kerong fish (*Terapon jarbua*) peptone using enzymatic hydrolysis

Y Srikandace*, S Priatni, S Pudjiraharti, W Kosasih and E S Endah

Research Unit for Clean Technology, Indonesian Institute of Sciences (LIPI), Komplek LIPI Bandung, Jalan Sangkuriang, Gedung 50, Bandung, 40135, Indonesia

*Email: yoice_srikandace@yahoo.co.id

Abstract. The Kerong fish (*Terapon jarbua*) peptone was produced by using alcalase and papain enzymes. The parameters of fish peptone development were temperature (50-65°C), pH 7, and hydrolysis time (4-7 h). Papain and alcalase were used to digest the substrate with concentrations of 0.2% (w/w). The results showed that the highest yield of peptone was resulted by using alcalase and papain enzymes at 65°C for 6 h and 5 h, respectively. The maximum harvest of peptone using alcalase enzyme was shown by soluble protein content of 6.382 mg/ml with N total of 0.535. The highest peptone production using papain enzyme revealed soluble protein content of 5.781 mg/ml with N total of 0.252. It indicated that alcalase enzyme produced more fish peptone *T. jarbua* than papain enzyme. The resultant fish peptone supported the growth of *Pseudomonas aeruginosa*, *Salmonella thypii*, *Bacillus subtilis* and *Clostridium sporogenes*.

1. Introduction

Nowadays, large amounts of protein-rich raw materials from fish processing are discarded without recovery. The wastes generated from fish processing industry are approximately 100,000 tons per year [1], including fish that cannot be used in industrial process or as side dishes [2]. In West Java-Indonesia, Kerong fish (*Terapon jarbua*) is abundant in population and less desirable because the taste of fish is not tasty. This type of fish is usually sold as low-valued products even though they contain proteins and essential amino acids [3,4]. The *T. jarbua* fish contains water soluble protein hydrolysate in mixture of polypeptides and amino acids such as peptone [5]. Peptone is a cheap material for microbial growth medium and eco-friendly. Peptone is a protein generated from fish with the addition of proteolytic enzymes in order to accelerate the hydrolysis process [6]. Several proteolytic enzymes are used to hydrolyze fish peptone such as alcalase, papain, bromelain, ficin, pepsin, trypsin, and pancreatin [1,3]. Alcalase and papain are low cost per unit of enzyme activity compared to other enzymes; these enzymes are used to prevent enzyme waste and to increase the commercial viability of the peptone product, especially for medium culture [7,8].

Preliminary study [5] showed that the highest yield of peptone from *T. jarbua* was achieved by 0.2% w/w of non-commercial papain enzyme hydrolysis at pH 7. Therefore, this present study aims to optimize the production of *T. jarbua* fish peptone using commercial alcalase and papain enzyme at varied temperature of 50-65°C, hydrolysis time (4-7 h), as well as to characterize the fish peptone product.

2. Methods

2.1. Fish peptone preparation
The fresh *T. jarbua* fish was originated from Indramayu-West Java. The commercial enzymes (papain XF 1601001 and alcalase XF 3612001) were purchased from Arisumcompiharm-a local manufacturer. A total of 100 g of fish was mixed with water with a ratio 1:4, blended and separated from fish bone. The fish porridge was then hydrolyzed at pH 7 with 0.2% w/w with commercial alcalase and papain enzymes, respectively. The enzymes were added into the mixture, heated at 50-65°C for 4-8 h and inactivated at 85°C for 15 min. The hydrolyzed mixtures were centrifuged (7000*g*, 20 min) at 4°C. Finally, the supernatant was collected and dried to obtain soluble peptone.

2.2. Assay for total protein
The total protein of *T. jarbua* fish peptone was carried out using Lowry's method. Lowry A solution consisted of 4 mg/mL of NaOH and 20 mg/mL of Na_2CO_3 in water. Lowry B solution consisted of 10 mg/mL of potassium sodium tartrate and 5 mg/ mL of $CuSO_4$ in water. Lowry C solution was prepared by mixing 50 mL of Lowry A solution and 1 mL of Lowry B solution. While Lowry D solution was prepared by diluting Folin-Ciocalteu reagent with distilled water (dH_2O) at equal volumes. As much as 500 µl of fish peptone samples were added with 2.5 mL of Lowry C solution and dissolved for 10 min. Samples were then added with 0.5 mL solution D and dissolved for 30 min. Sample solution was measured at 500 nm.

2.3. Assay for total N-amino
The total N-amino of *T. jarbua* fish peptone was estimated by formol titration method. A total of 10 mL sample was mixed with 20 mL of dH_2O, 40 mL of potassium oxalate and 3-4 drops of phenolpthalein indicator (PP). Sample was titrated using 0.1N NaOH, added with 2 mL of formalin and 3-4 drops of PP indicator. Sample solution was titrated using 0.1N NaOH and its protein content was measured.

2.4. Microbiology analysis
Bacteria *Escherichia coli, Staphylococcus aureus, Bacillus subtilis, Salmonella thypiimurium, Clostridium sprogenes, Enterobacter cloacae* and *Pseudomonas aeruginosa* were used for the growth test of fish peptone. Fish peptone with the highest yield was selected for this test. The sterilized broth test medium consisted of 1 g of fish peptone extract, 100 mL of dH_2O, 0.5% of yeast extract and 1% of NaCl at pH 7. The sterilized broth commercial medium contained 1 g of commercial peptone, 100 mL of dH_2O, 0.5% yeast extract and 1% NaCl at pH 7. Pure culture was regenerated into 100 mL of broth test media (conventional and commercial, respectively) and incubated at 37°C for 24 h and shaken at 150 rpm. Then, culture was incubated in Plate Count Agar (PCA). Total Plate Colony (TPC) method was applied to growth performance of fish peptone in PCA [6].

3. Result and discussion

3.1. Total protein and N-amino content
Analysis of Kerong fish (*T. jarbua*) was conducted to measure the protein and amino content of the raw material. The value of total protein and N-amino content from various temperatures and time incubations at pH 7 using commercial enzymes is described in Table 1.

Table 1 shows that the highest yield of fish peptone using papain enzyme and alcalase was obtained at incubation temperature 65°C. The highest yield of fish peptone that hydrolyzed with papain revealed total protein and N-amino were about 5.78 mg/mL and 0.35%, respectively. Hydrolysis with alcalase resulted in fish peptone with total protein and N-amino of 6.38 mg/mL and 0.54%, respectively. The research of fish peptone (protein content of 1.21%) from catfish (*Pangasius* sp) was obtained with the optimum temperature 60°C, 4% of concentration of papain for 48 h at pH 5 [7]. The fish peptone production using alcalase from ponyfish (*Eubleekeria splendens*), the yellow-stripped trevally

(*Selaroides leptolipis*) and mackerel (*Decapterus maruadsi*) were obtained at temperature 61.23°C, 6% alcalase concentration, and pH 8 in 27 min of hydrolysis [3]. The recent study showed that 0.2% papain and alcalase enzymes concentration at pH 7 produced high yield of fish peptone. The optimum yield was achieved might be due to different ability of the used enzyme to digest the fish substrate based on the temperature, enzyme concentration and hydrolysis time.

Table 1. Total protein and N-amino content from fish peptone at pH 7 using commercial enzyme.

Temperature	Incubation time (h)	Protein mg/mL		N-Amino (%)	
		Papain	Alcalase	Papain	Alcalase
50°C	4	1.69	2.21	0.08	0.43
	5	1.55	2.41	0.11	0.31
	6	1.66	2.54	0.24	0.38
	7	1.56	2.36	0.35	0.35
55°C	4	1.85	2.60	0.22	0.28
	5	2.08	2.70	0.21	0.25
	6	2.01	2.71	0.22	0.31
	7	2.01	2.28	0.18	0.38
60°C	4	4.30	2.42	0.17	0.25
	5	4.57	2.71	0.17	0.33
	6	4.43	3.13	0.17	0.34
	7	4.78	3.90	0.18	0.37
65°C*	4	5.45	5.99	0.25	0.50
	5	**5.78**	**6.41**	**0.35**	**0.58**
	6	4.80	6.38	0.30	0.54
	7	3.28	6.27	0.25	0.50

*Optimum hydrolysis condition for producing the highest yield of total protein and N-amino of fish peptone using commercial enzymes.

The chemical characteristic of fish peptone from *T. jarbua* showed the solubility and ratio of alpha-amino nitrogen (AN) to total nitrogen (TN) from fish peptone. Fish peptone resulted from hydrolysis with papain enzyme shows 92.17% of solubility and 6.06% of AN/TN. Fish peptone hydrolyzed by alcalase showed 95.11% of solubility and 8.48% of AN/TN. It indicates that fish peptone from *T. jarbua* can be used as culture media for bacteria. The study of silver carp filleting also showed fish peptone as a nitrogen source for growth medium of *S. aureus* [9]. The fish protein hydrolysate can be used as carbon and/or nitrogen source in microbial growth, which is not only an eco-friendly approach but can also produce cheaper raw materials for media culture formulation and other applications in food biotechnology [2,10].

3.2. Microbiology analysis

The growth of bacteria on peptone from 10^{-5} dilution stage conducted in PCA medium is shown in Table 2.

Table 2 and Figure 1 describe that all commercial peptone and fish peptone supported the growth of bacteria. Bacteria *E. coli, S. aureus, B. subtilis* and *P. aeruginosa* could grow well in 1x24 h of incubation but others more times. The growth also depended on the nutrition element support in media and show the specific color of colonies in particular medium [11,12].

Table 2. Total bacteria growth (cfu/mL).

Bacteria	Commercial peptone	Fish peptone resulted from hydrolysis with papain	Fish peptone resulted from hydrolysis with alcalase
Escherichia coli	71 x 10^{-5}	**181 x 10^{-5}**	56 x 10^{-5}
Pseudomonas aeruginosa	>200 x 10^{-5}	>200 x 10^{-5}	45 x 10^{-5}
Enterobacter cloacae	132 x 10^{-5}	26 x 10^{-5}	46 x 10^{-5}
Salmonella tyhpii	>200 x 10^{-5}	>200 x 10^{-5}	10 x 10^{-5}
Bacillus subtilis	>200 x 10^{-5}	32 x 10^{-5}	>200 x 10^{-5}
Stapylococcus aureus	114 x 10^{-5}	30 x 10^{-5}	14 x 10^{-5}
Clostridium sporogenes	81 x 10^{-5}	6 x 10^{-5}	>200 x 10^{-5}

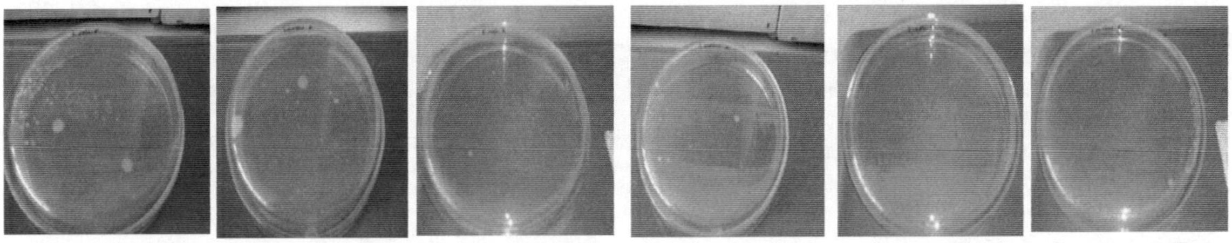

E. coli (P) *S. aureus* (P) *E. coli* (A) *S. aureus* (A) *E. coli* (Pn) *S. aureus* (Pn)

Figure 1. The colonies of bacteria cultured in commercial peptone and fish peptone.
P = Peptone hydrolyzed by papain, A = Peptone hydrolyzed by alcalase, Pn = Commercial peptone.

Figure 1 shows the growth of *E. coli* (negative gram bacteria) and *S. aureus* (positive gram bacteria) in PCA plate after all bacteria cultured in pepton. The result showed rapid growth of *E. coli, P. aeruginosa,* and *S. thypii* as negative gram bacteria using fish peptone hydrolyzed with papain enzyme. The total number of each bacteria were of 181x10^{-5}, >200x10^{-5} and >200x10^{-5}, respectively. On the other side, *B. subtilis* and *C. sporogenes* (positive gram bacteria) also showed rapid growth >200x10^{-5} cell on 10^{-5} dilution stage. In general, all bacteria grew well in commercial peptones. The interesting phenomena revealed from *E. cloacae* (negative gram bacteria) and *S. aureus* (positive gram bacteria). Bacteria *E. cloacae* and *S. aureus* were grown in peptones with papain and alcalase enzymes, respectively. Both bacteria did not grow well in the resultant peptone media, as shown by total colony of each bacteria of 26x10^{-5} and 14x10^{-5}, respectively. Previous study showed that the growth of *S. aureus* using fish peptone from *T. jarbua* was lower than that of commercial peptone [5]. It also indicated that fish peptone of *T. jarbua* was not suitable for the growth media of *E. cloacae* and *S. aureus*. Bioactive compounds from fish protein hydrolysates could contain base, amino acid, endogenous enzymes, bacteria, and digestive protease [13]. The present study also showed that peptone from *T. jarbua* could become specific growth medium for bacteria. It might replace expensive media for bacteria culture with the same quality.

4. Conclusion

Fish peptone as microbial growth media has been produced from *T. jarbua* through hydrolysis with commercial enzyme using an optimum condition of 0.2% papain enzyme concentration, pH 7 and 65°C for 5 h. Commercial alcalase enzyme produced more fish peptone *T. jarbua* than papain enzyme.

Acknowledgement
This study was funded by LIPI in 2016 and 2017. The authors also thanked to the collaboration of staffs and supply the material.

IOP Conf. Series: Earth and Environmental Science **160** (2018) 012009 doi:10.1088/1755-1315/160/1/012009

References

[1] Wisuthiphaet N, Kongruang S and Chamcheun C 2015 *J. Med. Bioeng.* **4** 466-70
[2] Husin N, Siti M M K, Ling T C, Nurul F M and Nohani J 2015 *IOP Conf. Biomed. Eng. Technol.* **81** 54-7
[3] Nicharee W, Sopa K and Sasithorn K 2016 *Int. J. Appl. Sci. Technol.* **9** 261-70.
[4] Ghaly A E, Ramakrishnan V V, Brook M S, Budge S M and Dave D 2013 *J. Microb. Biochem. Technol.* **5** 107-29
[4] Sabrah M M, Lamiaa I M, El-Sawy M and Emad H A E 2015 *J. Fish. Aquat. Sci.* **11** 147-62
[5] Srikandace Y, Priatni S, Pudjiraharti S, Wawan K and Lucia I 2017 *IOP Conf. Ser.: Earth Environ. Sci.* **60** 012005
[6] Sutton S 2011 *J. Validat. Technol.* **17** 42-7
[7] Utomo B S B, Suryanigrum T D and Harianto H R 2014 *Squalen Bull. Mar. Fisheries Postharvest Biotechnol.* **9** 115-26
[8] Murna M, Nofi S and Fahrizal Z 2012 *Int. J. Bioflux Soc.* **5** 36-40
[9] Meysam F, Somayeh B and Seyed R J 2015 *Food Sci. Nutr.* **3** 153-7
[10] Pere K, Mbatia B, Muge E and Wekesa V W 2017 *J. Appl. Biol. Biotechnol.* **5** 8-12
[11] Basu S, Bose C, Ojha N, Das N, Das J, Pal M and Khurana S 2015 *Bioinformation* **11** 182-4
[12] Shimizu K 2014 *Metabolites* **4** 1-35.
[13] Fernandes P 2016 *Front. Bioeng. Biotechnol.* **4** 59

Preparation and characterization of bacterial cellulose-beeswax films

Indriyati* **and L Indrarti**

Research Unit for Clean Technology, Indonesian Institute of Sciences (LIPI)
Komplek LIPI Bandung, Jalan Sangkuriang, Gedung 50, Bandung, 40135, Indonesia

*Email: indriyati@lipi.go.id

Abstract. Bacterial cellulose (BC) based polymer films containing carboxymethyl cellulose (CMC) and beeswax as additives were produced. Control film was prepared by mixing of BC slurry and CMC, while treated films were produced by the addition of beeswax/Tween 80 solution in various concentrations. The objective of this work was to study the effect of beeswax content on properties of BC films. The films were prepared by solution casting technique and dried in convection oven at 40 °C overnight. Addition of beeswax dissolved in Tween 80 affected mechanical properties, significantly reducing the tensile strength but increasing elongation of the control film by 22 and 43%, respectively. Beeswax+Tween80 addition also improved water solubility by doubled from 36 to 73-78%. However, there was no significant difference of WVTR and hydrophobic properties among those films with the addition of beeswax in Tween 80 into the mixtures of BC slurry+CMC.

1. Introduction

Cellulose is one of the most abundant polysaccharide in nature. It is the main constituent in the cell wall of vascular plants, where it is in complex with hemicelluloses, pectin, the aromatic polymer lignin and numerous glycoproteins [1]. In addition to plants, numerous microorganisms, such as green algae and oomycetes and various bacteria are able to produce cellulose as well. Bacteria *Acetobacter* strain can produce bacterial cellulose (BC), which has similar molecular formula with plant cellulose. However, BC is free of hemicellulose, pectin, lignin, and other biogenetic products associated with plant cellulose [2]. BC also exhibits remarkable properties such as high degree of polymerization [3], high water holding capacity [4], high crystallinity [5], and high mechanical strength [6]. Due to its biocompatibility and biodegradability, BC has been proposed for use in numerous medical applications [7-9] or as suitable material for edible films [10,11].

The main interest in edible films and coatings is generally based on their potential to control some properties, such as moisture, oxygen, flavor and/or aroma transfer, either between food components or to/from the atmosphere surrounding the food, with a resulting increase in quality and shelf-life of the food [12]. For edible film to be used in food area, it should have appropriate water barrier properties, good mechanical strength, low cost of raw materials, and simple technology for production [13]. BC is considered to be able to fulfill those requirements except for its water barrier property and flexibility. To overcome this drawback, BC can be combined with hydrophobic compounds, such as neutral lipids, fatty acids or waxes. Among lipids, beeswax (BW) has been successfully used to reduce the water vapor permeability (WVP) of biodegradable films [14,15] due to its high hydrophobicity and solid state at room temperature [16,17]. Moreover, it has been reported that BW could produce plasticizing effect to whey protein films [12]. BW is a mixture of many organic compounds, including wax esters, fatty acids, and hydrocarbons that exhibits viscoelastic behavior [18].

BC based films incorporated with BW have not been explored yet. This study will be another approach to improve biopolymer film properties for application in packaging field. Therefore, in the present study, effect of BW on the properties of BC-based films is investigated. Parameters such as thickness, solubility, water vapor transmission rate and water vapor permeability, and mechanical properties will be evaluated.

2. Experiment

2.1. Preparation of BC-based edible films

BC gels were purchased from local industry in Cianjur, West Java province, Indonesia. Washing, purification, and preparation of BC slurry were conducted as previously report [10]. A quantity of 200 g of BC slurry was casted on a non-sticky tray and dried overnight at 40 °C in a gear oven. The dried BC film was weighed and noted as the weight control for BC fiber. Another 200 g of BC slurry was stirred at 60 °C and added with carboxymethyl cellulose (CMC, Himedia). As the hydrophobic agent, BW (Cera Alba) with varied concentration was emulsified with 0.5 ml of Tween 80 (polyoxyethylene sorbitan monostearate, Merck). This solution was added into BC+CMC mixture and stirred until homogenous mixture was achieved. The blends were then degassed using a vacuum bell jar to remove the bubbles prior to cast onto the tray. Casting was conducted overnight at 40 °C in gear oven. BC based films were prepared according to formula in Table 1.

Table 1. Formulation of BC based films incorporated with BW.

Sample	BC slurry (g)	CMC (g)	BW (g)	Tween 80 (mL)
A	200	0.5	0	0.5
B	200	0.5	0.005	0.5
C	200	0.5	0.010	0.5
D	200	0.5	0.015	0.5

2.2. Characterization of BC-based edible films

Film thickness was measured before the WVP test using a digital micrometer (Mitutoyo, Japan), taking measurements at 3 random positions on the film. Averaged values of the thickness were used in all the WVP calculations. Solubility was calculated as the percentage of film dry matter dissolved during immersion in distilled water for 24 h. Pieces of film (2 cm x 2 cm) were cut from each film and dried at 105 °C in an oven for 5 h. Samples were weighed to the nearest 0.0001 g and noted as initial dry weight (w_o). Samples were then immersed in 50 ml distiller water for 24 h under constant stirring at 100 rpm. Undissolved film materials were filtered through nylon filter, followed by oven drying at 105 °C for 24 h. The samples were then weighed and noted as w_f. Samples were measured in triplicate and the percentage of total soluble matter (% solubility) was calculated as follow.

$$\%solubility = \frac{(w_o - w_f)}{wo} x100\% \tag{1}$$

The mechanical properties were characterized by Orientec UCT-5T universal testing with 100 kgf load cell according to ISO 527- 1993E standard method. The samples were cut by dumbbell cutter according to ISO 527-2 type 5A. Prior to measurement, the thickness of each specimen along the narrow area was measured using digital micrometer (Mitutoyo) and averaged. At least five specimens of each sample were measured and computerized calculated to obtain the average value. The specimens were mounted in the film-extension grips of the testing machine and stretched at a rate of 10 mm/min until breaking. The measurement was conducted at 23 °C and RH 50%.

Water vapor transmission rate (WVTR) of films was determined gravimetrically according to ASTM E96, modified by Gontard *et al* [19]. The test film was sealed to a glass cup having 16 cm^2 mouth areas and containing 10 g of dried silica gel (0% RH; 0 mmHg water vapor pressure). The cups were placed in a desiccator maintained at 100% RH (32.23 mmHg water vapor pressure) with distilled water. The water vapor transferred through the film and absorbed by the desiccant was determined from the weight gain of the silica gel recorded at various times. The cups were weighed using analytical balance at 24, 38, 72, and 96 h intervals. Changes in weight of the permeation cell were recorded to the nearest 0.0001 g. All tests were conducted in triplicate at room temperature. WVTR and WVP were determined using the following equations (Equation 1, 2):

$$WVTR = (\frac{\Delta m}{\Delta t A}) \quad (2)$$

$$WVP = WVTR(\frac{L}{\Delta p}) \quad (3)$$

where Δm/Δt is the weight gain per time (g/day), A is the exposed surface area of the film (m^2), L is the thickness of the film (mm), and Δp is the difference of partial pressure (32.23 mmHg) [19].

3. Result and discussion

3.1. Film formation

Films were slightly white in color with opaque appearance as a function of BW concentration (Figure 1). They were homogenous that indicates good compatibility and the film's forming capability among those materials. BW is hydrophobic, therefore surfactant is needed for water based formulation. In this work, Tween 80 served as non-ionic surfactant and used as an emulsifier for improving miscibility of BW in aqueous BC slurry due to its high hydrophilic lipophilic (HLB) value i.e. 15 [20]. Moreover, no air bubbles or holes were observed throughout the films since all bubbles have been removed by degassing techniques prior to casting the films. These air bubbles were a concern for film production, because they could produce a reduction in the tensile strength and other film properties [21]. Control film (sample A) made with BC and CMC was strong enough and slightly rigid when stretched manually by hand. Modified films with BW and Tween 80 (sample B, C, and D) were more opaque in appearance and flexible enough to be handled. The addition of BW affected the films thicknesses (Table 2). When the wax content increased, the higher molecular contact between cellulose (BC and CMC) and wax compound might weaken the aggregation forces of the polymer chain, making the matrix more open [22].

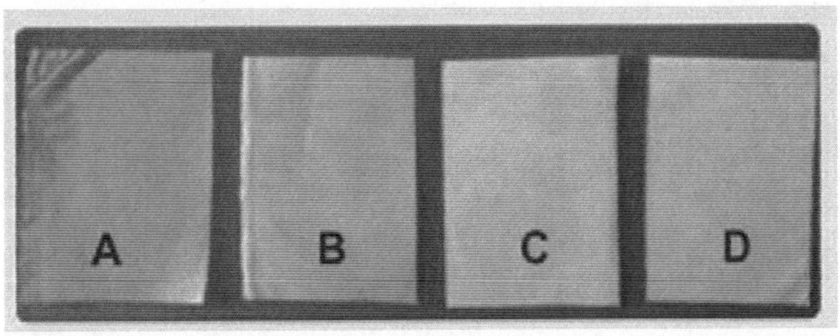

Figure 1. Physical appearance of BC based films without (A) and with (B, C, D) BW.

Table 2. Thickness and solubility of BC based films incorporated with BW.

Sample	Thickness (μm)	Solubility (%)
A	32.7 ± 6.8	33.5 ± 4.0
B	35.0 ± 4.0	78.8 ± 8.1
C	41.3 ± 2.3	79.5 ± 1.9
D	49.7 ± 8.1	66.6 ± 14.0

3.2. Water solubility

Film solubility in water is an important factor for edible applications. For example, films for water soluble pouches or edible strips must be readily soluble, while films on high-moisture foods must be insoluble. Table 2 shows water solubility of the films. BC films without BW were partially soluble in water of which its solubility 33.5±4.0% maintaining their integrity during immersion because of its rigid long chains and strong intra and inters hydrogen bonds [23]. Low percentage of solubility of BC-based edible films elucidates that only CMC was dissolved in water while the remaining solid were fragmented BC fibers. On the other hand, films with BW were highly soluble with its solubility ranging of 67-79% and broken apart in water after 24 h of immersion. This trend was in contrast to what have been reported previously [14] in which increasing the amount of BW decreased solubility. This indicates that the addition of BW into BC films weakened intra-molecular interactions in the aqueous condition compared to those of films without BW. The relatively high solubility of BC-BW films in water could possibly make the films appropriate for soluble pouches, like cellulose ether-based soluble pouch which is currently available commercially [24].

3.3. Mechanical properties of BC-based edible films

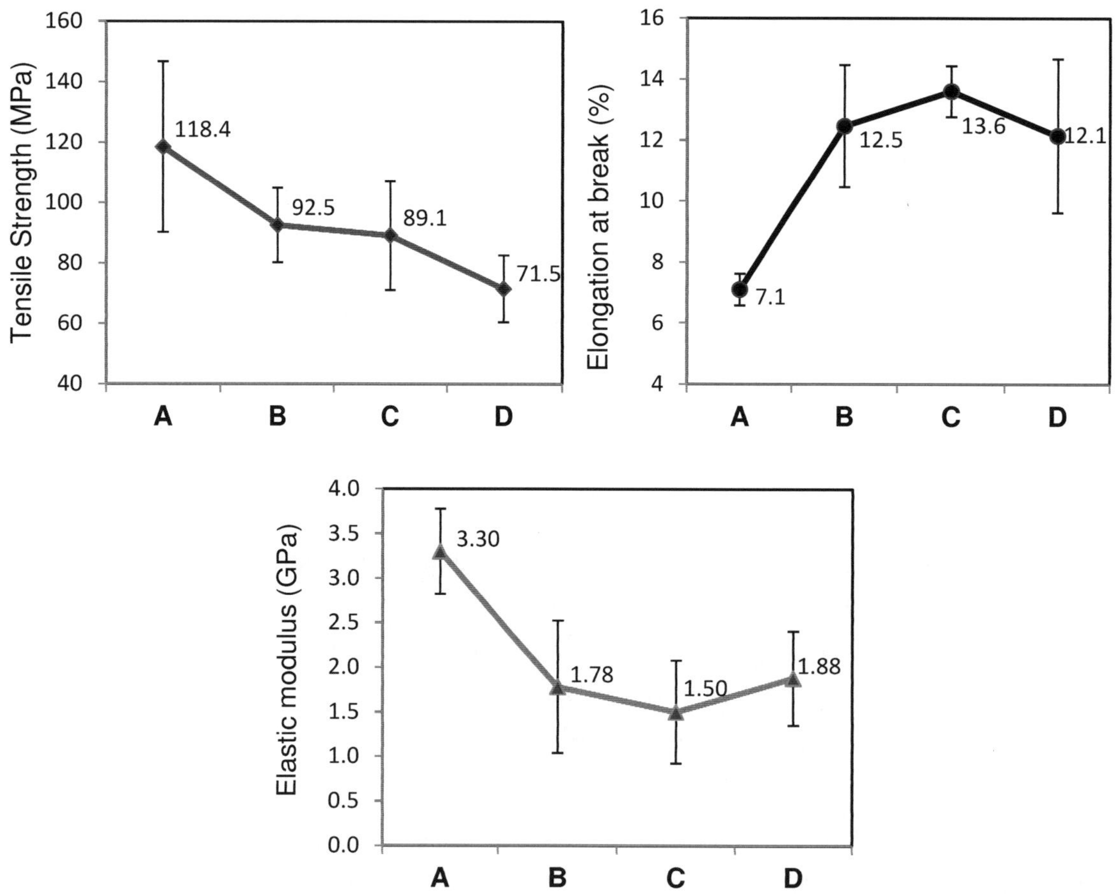

Figure 2. Mechanical properties of BC-based edible films with variation of beeswax.

Tensile properties are essential for edible films and coatings. It relates to durability as well as coating ability to increase mechanical integrity of foods [12]. High tensile strength (TS) is necessary for edible film in order to withstand the normal stress encountered during the application, shipping, and food handling. However, flexibility of edible films, i.e. elongation at break (EB) is also required for application of edible films. Elastic modulus (EM) is a measure of the intrinsic stiffness, durability or resilience of a film.

Comparison of mechanical properties of BC based films, with and without BW is depicted in Figure 2. TS of control film (BC+CMC) was 118.4±28.2 MPa. This value is comparable to our previous report [25]. It was observed that TS and EM of films were significantly decreased by the incorporation of BW. TS decreased by 22%, while EM decreased by 46%. Observed decrease with addition of BW might be due to the increase of free volume between polymer chains as reported earlier for starch films by Brandelero *et al* [26]. On the other hand, addition of BW remarkably increased the flexibility of films as shown by their EB values. Thus, it appears that BW had a plasticizing effect on BC based film. The presence of a significant amount of free fatty acids in BW may increase chain mobility and contribute to the flexibility of films. This result is in agreement with previous study, which demonstrated the plasticizing effect of BW in whey protein films [12]. These findings also support those of Tanaka *et al* [27] who studied the effect of different types of lipid materials on fish water soluble proteins (FWSP) composites films. They reported that the incorporation of BW remarkably increased the flexibility of films. Increasing concentration of BW at the same content of Tween 80 did not change the mechanical properties of BC based films as shown in B, C, and D films.

3.4. Water vapor transmission rate (WVTR) and water vapor permeability (WVP)

WVTR is the rate of water vapor permeating through the film, which is determined from the slope of the regression line of sample weight versus time graph and then divided by the area of the film being exposed to the transmission (Equation 2). Table 3 tabulates the WVTR values as well as WVP values calculated from the WVTR values (Equation 3). Table 3 shows that WVTR increased with an increase in the level of beeswax concentration. The similar trend was also observed for WVP values. These results were unexpected, because the addition of BW to the BC based film was supposed to reduce the transmission of water due to the hydrophobicity of BW. Study published by Auras *et al* [21] with cassava starch films containing glycerol and BW also documented this effect. They explained the phenomenon might be due to the lack of interaction of the lipid with the matrix, in our case with BC. This incompatible mix increased the chance for water molecules to migrate through the films through the BC/CMC/BW interface.

Table 3. The slope of the linear regression line of the sample weight vs time graph, the WVTR, and WVP of the BC based films with and without beeswax.

Sample	Slope, $\Delta W/\Delta t$ (g/day)	WVTR ((g/day)/m^2)	WVP (g.mm/(m^2.day.mmHg))
A	0.0072	4.43	0.00433
B	0.0073	4.46	0.00484
C	0.0088	5.36	0.00687
D	0.0093	5.75	0.00886

4. Conclusion

BC based films with incorporating BW were successfully produced with a stable film structure. The preparation technique was very simple through casting method. The addition of beeswax improved the water solubility of films from 50 to 63-83%. It also influenced mechanical properties of films, significantly reducing the tensile strength and elastic modulus as well as increasing elongation. However, different concentration of BW did not affect much on mechanical properties of films. Addition of 0.005 g of BW showed the lowest WVTR of BC-BW film.

IOP Conf. Series: Earth and Environmental Science **160** (2018) 012010 doi:10.1088/1755-1315/160/1/012010

Acknowledgement
The authors would like to acknowledge Indonesian Institute of Sciences through Riset Unggulan 2016 for financial support.

References
[1] Somerville C 2006 *Annu. Rev. Cell Dev. Biol.* **22** 53-78
[2] Brown A J 1886 *J. Chem. Soc. Trans.* **49** 432–9
[3] Dahman Y, Jayasuriya K E and Kalis M 2010 *Appl. Biochem. Biotechnol.* **162** 1647–59
[4] Saibuatong O A and Phisalaphong M 2010 *Carbohyd. Polym.* **79** 455–60
[5] Keshk S M A S 2014 *J. Bioproces. Biotech.* **4** 150
[6] Castro C, Zuluaga R, Putaux J L, Caro G, Mondragon I and Gañán P 2011 *Carbohyd. Polym.* **84** 96–102
[7] Lin W C, Lien C C, Yeh H J, Yu C M and Hsu S H 2013 *Carbohydr. Polym.* **94** 603-11
[8] Czaja W K, Young D J, Kawecki M and Brown R M 2007 *Biomacromolecules* **8** 1-12
[9] Klemm D, Schumann D, Udhardt U and Marsch S 2001 *Prog. Polym. Sci.* **26** 1561-603
[10] Indrarti L, Indriyati, Syampurwadi A and Pujiastuti S 2016 *AIP Conference Proceedings* **1711** 050007
[11] Tang X, Kumar P, Alavi S and Sandeep K 2012 *Crit. Rev. Food Sci.* **52** 426–42
[12] Talens P and Krochta J M 2005 *J. Food Sci.* **70** E239-E243
[13] Debeaufort F, Quezada-Gallo J A and Voilley A 1998 *Crit. Rev. Food Sci.* **38** 299-313
[14] Soazo M, Rubiolo A C and Verdini R A 2011 *Food Hydrocolloids* **25** 1251-5
[15] Monedero F M, Fabra M J, Talens P and Chiralt A 2009 *J. Food Eng.* **91** 509-15
[16] Yang L and Paulson A T 2000 *Food Res. Int.* **33** 571–8
[17] Morillon V, Debeaufort F, Blond G, Capelle M and Voilley A 2002 *Crit. Rev. Food Sci. Nutr.* **42(1)** 67–89
[18] Shellhammer TH and Krochta JM 1997 *J. Food Sci.* **62** 390–4
[19] Gontard N, Guilbert S and Cuq JL 1992 *J. Food Sci.* **57** 190–5
[20] Lo YL 2003 *J. Control. Release* **90** 37–48
[21] Auras R, Arroyo B and Selke S 2009 *Starch* **61** 463-71
[22] Hromis N M, Lazic V L, Markov S L, Vastag Z G, Popovic S Z, Suput D Z, Dzinic N R, Velicanski A S and Popovic L M 2015 *J. Food Eng.* **158** 86-93
[23] Pandey M, Abeer M M and Amin M C I 2014 *Int. J. Pharm. Pharm. Sci.* **6** 89-93
[24] Bamdad F, Goli A H and Kadivar M 2006 *Food Res. Int.* **39** 106-11
[25] Indrarti L and Indriyati 2017 *IOP Conf. Ser.: Earth Environ. Sci.* **60** 012018
[26] Brandelero RPH, Yamashita F and Grossmann MVE 2010 *Carbohydr. Polym.* **82** 1102–09
[27] Tanaka M, Ishizaki S, Suzuki T and Takai R 2001 *J. Tokyo Univ. Fish.* **87** 31-7

IOP Conf. Series: Earth and Environmental Science **160** (2018) 012011 doi:10.1088/1755-1315/160/1/012011

Antimicrobial activity of extracellular liquid obtained from molasses fermentation by *Nocardia* sp strain V1

P C Mawarda*, E S Endah, D Ratnaningrum, T A Budiwati, D G S Andayani and S Pudjiraharti

Research Unit for Clean Technology, Indonesian Institute of Sciences (LIPI) Komplek LIPI Bandung, Jalan Sangkuriang, Gedung 50, Bandung 40135, Indonesia

*Email: panji.cahya.mawarda@lipi.go.id

Abstract. Molasses is one of agricultural wastes that could be utilized as a substrate for microbial fermentation. The utilization of molasses as a fermentation medium is expected to increase its added value. In this study, molasses served as a carbon source for bolstering the growth and the production of extracellular liquid having antimicrobial activity produced by *Nocardia* sp strain V1. The six types of fermentation media containing three carbon sources namely molasses, molasses-starch, and molasses-glucose were prepared; wherein molasses used were the ones with and without the pre-treatment using $K_4Fe(CN)_6.3H_2O$. This chemical was used to precipitate heavy metals and other substances in molasses that might inhibit the growth of the bacteria. The antimicrobial activity of extracellular liquid produced by *Nocardia* sp strain V1 was evaluated against 10 microbes isolated from infected strawberries and pseudostem of banana plant. The result showed that the antimicrobial activities were demonstrated by the extracellular liquid derived from the bacteria grown in the media containing pre-treated molasses-starch (M2 medium), pre-treated molasses-glucose (M3 medium), and glucose-starch (synthetic medium) as the carbon sources. The extracellular liquid from those fermented media was able to inhibit 2 microbes from a total of 10 microbes identified as *Rhodotorula* sp and *Fusarium graminearum* with the inhibition diameter ranged from 10-21 mm.

1. Introduction

The attempts to harness byproducts from the sugar industry continue to expand along with the emergence of various products with high added value such as alcohol, artificial sweeteners, vitamins, amino acids, and so forth. The products converted from these by-products are widely used by the public in the field of food, pharmaceutical, and bioenergy [1]. One of the most common by-products produced in sugar industry is molasses. This compound is produced in the final stage of the sugar crystallization process and could potentially cause serious environmental problems. Molasses is known as one of the major sources of contaminants in the aquatic environment since it is brownish, has strong acid properties and high COD and BOD level [2,3]. Meanwhile, the production of molasses in Indonesia has been increasing since 1964 with a total of 1.36 million tons in 2013 [4]. This increasing number has encouraged the researchers to continue their efforts in utilizing and converting molasses into different products with high added value.

One of the applications of molasses utilization in the field of biotechnology is its use as a fermentation medium. This usage could reduce the production costs since the synthetic medium accounts for 30% of the total cost in the fermentation process [5]. The use of synthetic carbon sources such as glucose, sucrose, mannitol, fructose, and arabitol are not economical due to their high price. Therefore, finding the new and inexpensive carbon sources is an important aspect to obtain more cost-effective cultures [6,7,8]. Besides being affordable, molasses contains various nutrients including carbohydrates, vitamins, minerals, and other micronutrients needed for microbial growth. Carbohydrate contained in molasses comprises glucose and fructose with the fermentable total sugar ranges from 45-55% [1,4].

The use of molasses as a medium in various fermentation processes has been widely studied. Research conducted by Zhu *et al* found that molasses could serve as a fermentation medium for *Trichosporon fermentans* to produce biodiesel [9]. Molasses has also been used as a fermentation medium for producing lactic acid [10], polyhydroxybutyrate [11], ethanol [12], pullulan [13], cellulose [14], and xantan sap [15]. In this study, molasses was used as a fermentation medium for *Nocardia sp* strain V1 to produce extracellular liquid having antimicrobial activity.

Bacteria of the genus *Nocardia* has been shown to produce a variety of secondary metabolites that actively inhibit the microbial growth. For example, *Nocardia* sp strain WW-12651 was known to produce a broad-spectrum of antibiotic compounds from thiazolyl peptide group such as nocatin I, II and III. These compounds had the ability to inhibit the growth of gram-positive bacteria such as methicillin-resistant *Staphylococcus aureus* (MRSA), multidrug-resistant *Enterococcus faecium* (MREF), and penicillin-resistant *Streptococcus pneumoniae* [16]. *Nocardia* sp strain CS682 also produced an antibiotic compound called nargecin that could inhibit the growth of MRSA. In addition, other antibiotic compounds such as peptidolipin B and peptidolipin E were also produced by *Nocardia araeonsis* strain W9705 and proven to inhibit the growth of MRSA [17].

As regard to the potential of *Nocardia* as a source of antibiotic compounds, this study aimed to investigate whether *Nocardia* sp strain V1 was capable of producing extracellular liquid having antimicrobial activity. Extracellular liquid was obtained through fermentation with a variety of media composition containing molasses. The antimicrobial activity of this extracellular liquid was intended as a biocontrol agent that could be used in the fruits plantation. Biocontrol agents in the form of microorganism could serve as the alternative tools to replace chemical pesticides and fertilizers which in turn reduce the deleterious environmental impact of agriculture. Hence, the antimicrobial activity was evaluated against 11 microbial species isolated from infected strawberries and pseudostem of banana plant.

2. Methods

2.1. Microorganism and reagents

Nocardia sp strain V1 was the collection of Research Unit for Clean Technology, Indonesian Institute of Sciences. Molasses was obtained from the local sugar industry. Nutrient agar (NA), potato dextrose agar (PDA), glucose, starch, triptone, $CaCO_3$, PDA, and $K_4Fe(CN)_6.3H_2O$ were purchased from Merck (Darmstadt, Germany). Yeast extract was purchased from Laboratorios conda S.A (Madrid, Spain). The analytical grade of Nelson reagents was purchased from Merck (Darmstadt, Germany) and used for determination of total sugar concentration using Nelson-Somogyi method.

2.2. The isolation and identification of allegedly phytopathogenic microorganisms

The infected strawberries and pseudostem of banana plant were sampled aseptically from the local farm. Each sample was chopped, mixed with distilled water, and vortexed for 5 min prior to serial dilution. The samples were serially diluted in the ten-fold dilution method until the 10^{-5} dilution. Each serially diluted sample of 0.1 mL was spread on NA and PDA. The plates were incubated at 30°C for 48 h. The microorganisms were isolated and stored at 4°C. Some of the isolates were identified based on 18s rRNA in the School of Life Sciences, Bandung Institute of Technology.

2.3. Molasses pre-treatment

Molasses of 500 g was mixed with 500 mL distilled water and centrifuged at 7000 rpm for 15 min. The supernatant was collected and mixed with 3 mL of $K_4Fe(CN)_6.3H_2O$ to remove the heavy metals and other inhibitory substances. The mixture was heated at 70°C for 30 min and stored overnight. The mixture was centrifuged at 7000 rpm for 15 min and the supernatant was collected [18].

2.4. Molasses fermentation by Nocardia *sp strain V1*

The inoculum was prepared by inoculating *Nocardia* sp strain V1 into 250 mL medium containing glucose (10 g/L), starch (10 g/L), triptone (5 g/L), yeast extract (2.5 g/L), and $CaCO_3$ (1 g/L) at pH 7. The medium was incubated in a shaker incubator at 30°C with agitation velocity of 150 rpm for 48 h. The inoculum of 10% (v/v) as inoculated into 250 mL fermentation media containing different compositions. The seven types of fermentation media were prepared as follows:

a. M1 (pre-treated molasses (20 g/L), triptone (5 g/L), yeast extract (2.5 g/L), and $CaCO_3$ (1 g/L);
b. M2 (pre-treated molasses (20 g/L), starch (5 g/L), triptone (5 g/L), yeast extract (2.5 g/L), and $CaCO_3$ (1 g/L);
c. M3 (pre-treated molasses (20 g/L), glucose (5 g/L), triptone (5 g/L), yeast extract (2.5 g/L), and $CaCO_3$ (1 g/L);
d. M1a (untreated molasses (20 g/L), triptone (5 g/L), yeast extract (2.5 g/L), and $CaCO_3$ (1 g/L);
e. M2a (untreated molasses (20 g/L), starch (5 g/L), triptone (5 g/L), yeast extract (2.5 g/L), and $CaCO_3$ (1 g/L);
f. M3a (untreated molasses (20 g/L), glucose (5 g/L), triptone (5 g/L), yeast extract (2.5 g/L), and $CaCO_3$ (1 g/L).
g. Synthetic Medium (glucose (10 g/L), starch (10 g/L), triptone (5 g/L), yeast extract (2.5 g/L), and $CaCO_3$ (1 g/L).

The pH of each medium was 7.0. The inoculated media were shaken at 150 rpm and incubated at 30°C for 7 days. The fermented medium of 10 mL was sampled each day and centrifuged at 7000 rpm for 15 min to collect the supernatant.

2.5. Antimicrobial activity test

The collected supernatant from each fermented medium was subjected to antimicrobial activity test using agar diffusion method. The solid media containing PDA which previously mixed with 1 mL of each microbial suspension were prepared in a sterile petri dish. The supernatant of 20 μL was dropped into a sterile paper disc previously put in the media and incubated for 72 h at 30°C. The observations were made by measuring the diameter of the inhibition zone in millimeters.

2.6. Determination of total sugar consumption rate and pH measurement [19]

The determination of total sugar concentration in the sample was determined as follows: the supernatant of 1 mL was collected after 7 days of incubation. The sample was subjected to strong acid hydrolysis by mixing with 0.5 mL of concentrated H_2SO_4. The mixture was incubated at 100°C for 1 h. Afterwards, the mixture was diluted to 10 mL distilled water. The diluted sample of 1 mL was added into the test tube and mixed with 1 mL Nelson reagent. The mixture was incubated for 20 min in a 100°C water bath. The sample was cooled down, mixed with 1 mL arsenomolybdate solution, and added with 7 mL distilled water. The solution mixture was mixed by vortex and subjected to absorbance measurement using a spectrophotometer at 520 nm. The absorbance value was plotted into the standard curve to obtain the total sugar concentration on each sample. The percentage of total sugar consumption was calculated by comparing with the initial sugar concentration from each medium before inoculation obtained through the same method. The pH of each fermented medium was measured every day by a pH meter to monitor the acidity of the medium during fermentation.

3. Results and Discussion

3.1. The isolation and identification of allegedly phytopathogenic microbes

The microbes were isolated from infected strawberries and pseudostem of banana plant in order to obtain allegedly phytopathogenic microbes that could be inhibited by the extracellular liquid being produced.

The result showed different colonies of microbes appeared on PDA in the dilution of 10^{-1}, 10^{-2}, 10^{-3}, and 10^{-4}. However, the microbial colonies did not appear on NA. This indicated that the microbes infecting strawberries and banana's pseudostem were fungi, yeasts, and molds. These microorganisms prefer PDA to NA since media containing high carbohydrate and nitrogen source are required for the growth of fungi [20]. PDA contains the infusion of potato and dextrose that foster luxuriant fungal growth whereas NA offers the cultivation of microbes supporting the growth of a wide range of non-fastidious microorganisms and contains many nutrients needed for the bacterial growth [21]. The result showed six microbes suspected as fungi were obtained from the infected banana's pseudostem and coded as Cador, Caue, Cadung, Cabe1, Cadom, Caug and 1 fungus was identified as *Fusarium graminearum* based on 18s rRNA analysis (Figure 1).

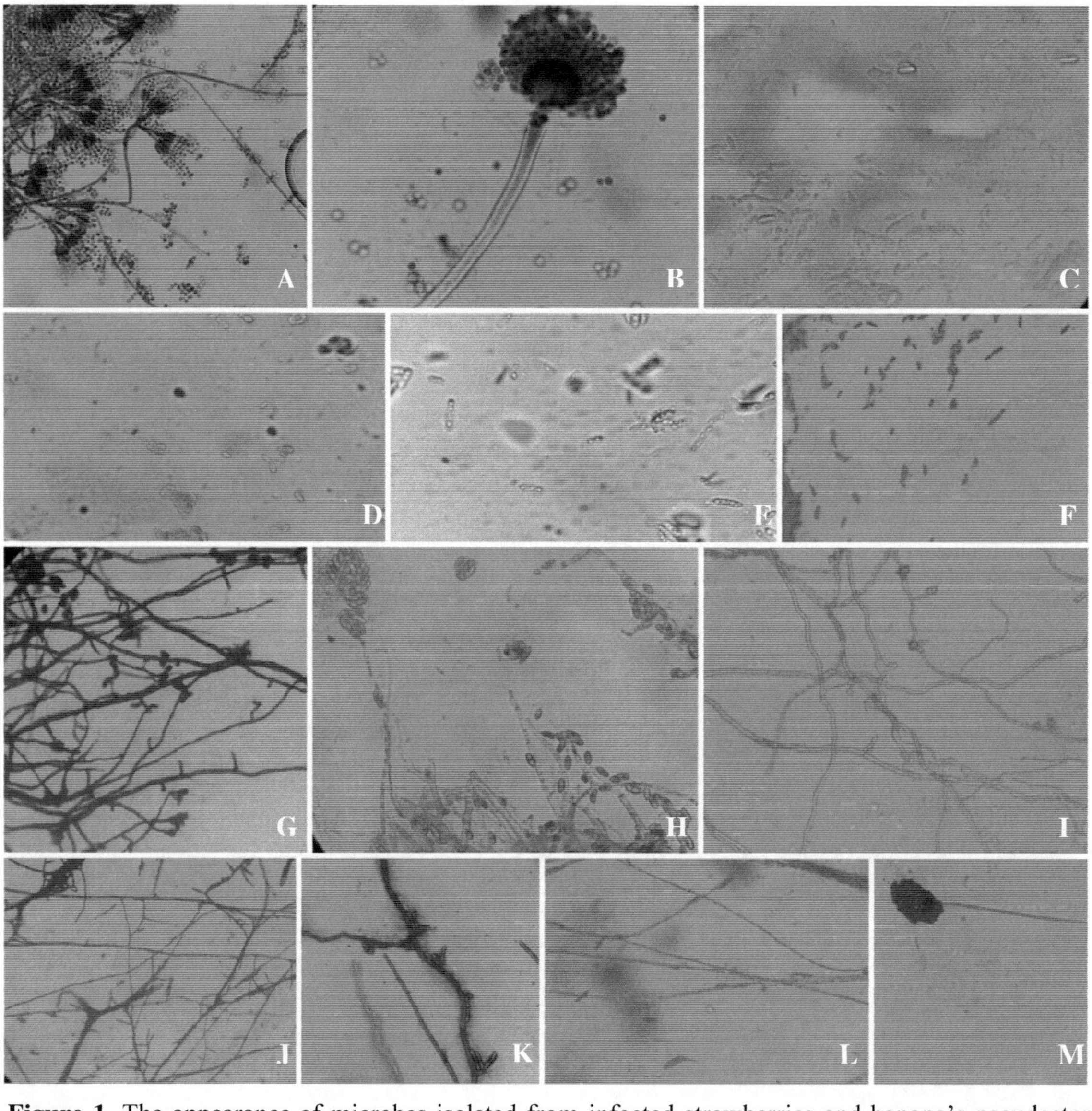

Figure 1. The appearance of microbes isolated from infected strawberries and banana's pseudostem under a microscope with magnification of 40x10 (A: *Aspergillus sp,* B: *Penicillium sp,* C: *Rhodotorula sp,* D: *Candida psudotropicalis,* E: *Kluyveromyces sp,* F: *Candida krusei,* G: *Fusarium graminearum,* H: Cador, I: Caue, J: Cadung, K: Cabe1, L: Cadom, and M: Caug).

Fusarium graminearum has been identified as a pathogen that causes devastating disease in banana and head blight of small grain cereals including wheat and barley [22,23]. This fungus spreads the spores on or inside flowering spikelets [24]. The infection of this fungus in crops causes great economic loss due to the reduced quantity and quality as well as to the contamination by diverse mycotoxins, including deoxynivalenol (DON) and zearalenone (ZEA), which are harmful for humans and animals [24]. DON is an effector molecule that inhibits protein synthesis through the activation of activating critical cellular kinases involved in signal transduction and binding to the ribosome disrupting growth, immune function, and reproduction [25]. Meanwhile, ZEA has hepatotoxic, haematotoxic, immunotoxic and genotoxic properties due to the formation of two metabolites a-zearalenol (a-ZEA) and b-zearalenol (b-ZEA) [26].

In addition, two fungi and four yeasts were successfully isolated from infected strawberries and identified based on 18s rRNA analysis as *Aspergillus* sp, *Penicillium* sp, *Rhodotorula* sp, *Candida psudotropicalis, Kluyveromyces* sp, and *Candida krusei* (see Figure 1) [27]. Along with microbes isolated from infected banana's pseudostem, the microbes having significant influence to cause strawberry diseases were subjected to antimicrobial activity test. They were *Aspergillus* sp, *Penicillium* sp, and *Rhodotorula* sp. *Aspergillus* species are known to be responsible for several disorders in various plants and plant products including strawberries. They contaminate agricultural products during pre-harvest, harvest, processing and handling stages [28]. *Aspergillus* causes root diseases in strawberry and responsible for its post-harvest decay. This microbe also produces mycotoxins such as aflatoxins and ochratoxin A that are responsible for carcinogenicity and immunotoxicity [29]. Aside from *Aspergillus, Penicillium* is also one of the important microbes causing diseases in strawberries. *Penicillium* species are one of the main pathogenic fungi that decay strawberry fruit and cause fruit blotch disease [30]. Among the fungi isolated from strawberries, *Penicillium* spp are also known as potential mycotoxin producers [31, 32]. They are the most common fungal genus on the surface of the strawberries and the most frequently isolated genus growing from both conventional and organic grown berries [33]. In addition to *Aspergillus* and *Penicillium, Rhodotorula* is also found as the most common yeast isolated from strawberries [30]. *Rhodotorula* spp could cause skin, ocular, meningeal, prosthetic joint, and peritoneal infections in human [34]. A study conducted by Tomsíková showed the contamination of *Rhodotorula* sp in food that was provided for immunocompromised patients in hospitals [35]. Due to its severity and pathogenicity, these three microbes along with other seven microbes isolated from infected banana's pseudostem were tested against extracellular fluid produced by *Nocardia* sp strain V1 through molasses fermentation.

3.2. Antimicrobial activity

The medium composition was optimized to determine which medium favored *Nocardia* sp strain V1 to produce the highest antimicrobial activity. The molasses pre-treatment through the addition of $K_4Fe(CN)_6.3H_2O$ was also investigated to know whether this treatment had significant influence on the production of antimicrobial extracellular liquid. The pre-treatment of molasses is required since it contains a lot of suspended particles and complex structures which cause heterogeneity in the medium and affect the cell growth rate. Therefore, many types of molasses treatment have been proposed in order to remove undesirable substances such as coloring substances, heavy metals, and unknown compounds that may act as growth inhibitors [36]. This could be done by heat treatment, acidification treatment, and metal complexing agent treatment [37].

The result exhibited *Nocardia* sp strain V1 grown in the media containing pre-treated molasses-starch (M2) and pre-treated molasses-glucose (M3) produced extracellular liquid having antimicrobial activity (see Table 1). Meanwhile, *Nocardia* sp strain V1 grown in all media containing untreated molasses did not show any inhibition on the growth of allegedly phytopathogenic microbes. $K_4Fe(CN)_6.3H_2O$ acted to reduce the load of various fungal growth inhibitory substances in molasses like Ca, Cu, and Mn. The chemical served as metal complexing agent that reacted with those heavy metals, causing their precipitation. This compound was proven capable of complexing 18 out of 21 microelements found in molasses [38]. Hence, the growth of *Nocardia* sp strain V1 in the media containing pre-treated molasses was better since the inhibitory substances were removed by

$K_4Fe(CN)_6.3H_2O$. This led to better production of extracellular liquid that could inhibit the growth of allegedly phytopathogenic microbes. Some research papers have shown that the molasses pre-treatment using $K_4Fe(CN)_6.3H_2O$ also helped to increase the production of inulinase [39], pullulan [40], and succinic acid [41]. In contrast, pre-treated molasses did not increase the production of lactic acid indicating that molasses contained desirable substances which significantly affected lactic acid production [10].

Table 1. The antimicrobial activity of extracellular liquid produced by *Nocardia* sp strain V1 in different fermentation medium.

Allegedly phytopathogenic microbes	Medium type	Diameter of inhibition zone (mm)			
		Day-4	Day-5	Day-6	Day-7
Fusarium graminearum	M1	-	-	-	-
	M2	-	-	21	20
	M3	-	-	20	17
	M1a	-	-	-	-
	M2a	-	-	-	-
	M3a	-	-	-	-
	Synthetic medium	15	18	-	-
Rhodotorulla sp	M1	-	-	-	-
	M2	-	10	14	10
	M3	-	10	14	12
	M1a	-	-	-	-
	M2a	-	-	-	-
	M3a	-	-	-	-
	Synthetic medium	15	20	-	-
Cador, Caue, Cadung, Cabe1, Cadom, Caug, *Aspergillus* sp, *Penicillium* sp	M1	-	-	-	-
	M2	-	-	-	-
	M3	-	-	-	-
	M1a	-	-	-	-
	M2a	-	-	-	-
	M3a	-	-	-	-
	Synthetic medium	-	-	-	-

Nocardia sp strain V1 grown on the synthetic, M2, and M3 media produced extracellular liquid that was active to inhibit 2 out of 10 allegedly phytopathogenic microbes identified as *Fusarium graminearum* and *Rhodotorulla* sp (see Table 1). The inhibition zone diameter varied depending on the fermentation time, medium type, and allegedly phytopatogenic microbes being inhibited. When the extracellular liquid was tested against *Fusarium graminearum*, the highest antimicrobial activity was shown by M2 and M3 media on the sixth day of fermentation, displaying the widest inhibition zone diameters of 21 mm and 20 mm respectively. Meanwhile, the extracellular liquid from the synthetic medium showed less ability to inhibit *Fusarium graminearum* with inhibition zone diameters of 15 mm and 18 mm on the fourth day and the fifth day of fermentation respectively. However, the highest antimicrobial activity against *Rhodotorulla* sp was shown by extracellular liquid obtained from the synthetic medium. This was also appeared on the fifth day of fermentation with the inhibition zone diameter of 20 mm. Meanwhile, the M2 and M3 media formed their widest inhibition zone with a diameter of 14 mm each when they were tested against *Rhodotorulla* sp. This appeared on the sixth day of fermentation. Nonetheless, the bacteria grown in the medium containing molasses alone as a carbon source did not show any antimicrobial activity. This might be due to the absence of glucose and starch as carbon sources used in the starter culture, incommoding the bacteria to adapt in a new environment and hampering them from producing antimicrobial metabolites. Similar result was shown by Kotzamanidis *et al* [10] who revealed that the highest lactic acid concentration was produced by the bacteria grown in the medium containing molasses mixed with sucrose and yeast extract. They suggested that sucrose and yeast extract played a key role to boost up the concentration of lactic acid in the culture whereas molasses did not.

The result also showed that *Nocardia* sp strain V1 produced their maximum inhibition zone in different fermentation time depending on the medium composition. In the synthetic medium, the bacterium could utilize the nutrients faster since they were less complex compared to the ones containing molasses. Molasses contains various complex chemicals in addition to sugars such as waxes, sterols, phospholipids, non-nitrogenous compounds, and ash contents [33]. This made the bacteria grew longer due to slow adaptation process in the presence of those chemicals and confined the early production of antimicrobial metabolites than those grown in the synthetic medium. Therefore, *Nocardia* sp strain V1 grown in the synthetic medium was able to inhibit the allegedly phytopathogenic microbes earlier on the fourth day of fermentation whereas the bacteria grown in the medium containing molasses inhibited later. The antimicrobial activity of extracellular liquids in the M2 and M3 media decreased on the seventh day of fermentation while the ones in synthetic medium weakened on day sixth. This declining trend was possibly caused by the reduction of available nutrients inside the medium leading to the drop of bacterial growth which in turn slowed down their metabolism and decreased the production of antimicrobial metabolites.

3.3. Total sugar concentration and pH measurement

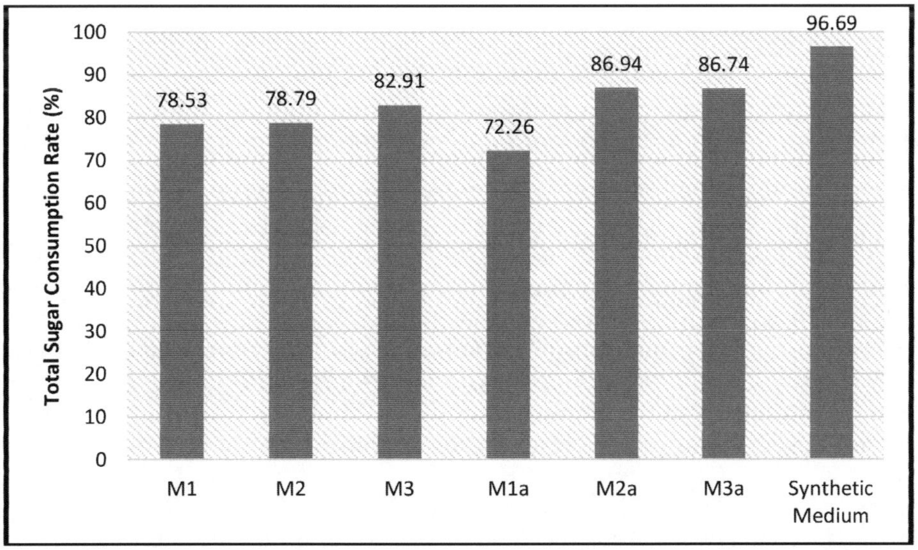

Figure 2. Total sugar consumption of the bacteria grown on various media.

The total sugar concentration was measured during fermentation time to determine the sugar consumption rate on each medium. The data could imply the ability of the bacteria grown on each medium to produce the extracellular metabolites having antimicrobial activity. The high sugar consumption rate indicated the high nutrient utilization by the bacteria to undergo their metabolism. The result showed the highest sugar consumption rate of 96.68% belonged to the bacteria grown in the synthetic medium (see Figure 2). This data indicated that the bacteria grown in the synthetic medium had the highest efficiency to utilize the sugar for their metabolism which was confirmed by the higher antimicrobial activity to inhibit *Rhodotorula* sp compared to the others.

The bacteria grown in the media containing molasses-starch and molasses-glucose showed higher sugar consumption rate compared to the ones containing molasses alone regardless the pre-treatment of molasses (see Figure 2). The bacteria grown on M2 and M3 media had sugar consumption rate of 78.80% and 82.91% respectively, being higher than the media containing pre-treated molasses alone which was 78.50%. This was also confirmed by the fact that the antimicrobial activity of the bacteria grown in those media was higher when they were tested against *Fusarium graminearum* compared to the others. This trend was similar to the media containing untreated molasses (see Table 1). The sugar consumption rate of the bacteria grown on M2a and M3a media were 86.94% and 86.73% respectively, being higher compared to the one containing untreated molasses alone which was 72.26%. However, they did not show any antimicrobial activity since the inhibitory compounds in molasses were not removed.

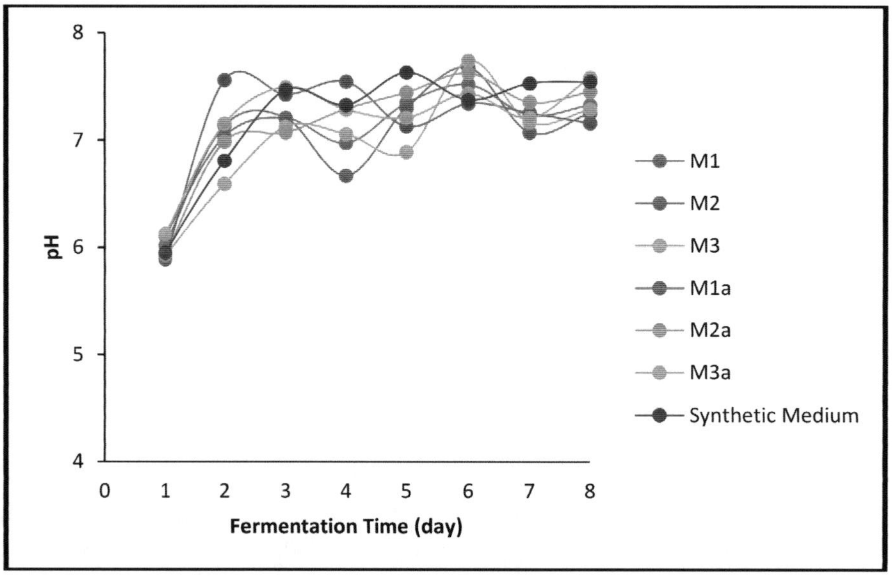

Figure 3. The change of pH on each medium during fermentation.

The pH of each fermented medium increased as fermentation time took longer. *Nocardia* sp strain V1 employed in this study actively fermented all the media and increased the pH to as high as 7.58 which belonged to those grown in the synthetic medium. As shown in Figure 3, the pH of each fermented medium was similar to each other since the acidity of the resulting metabolites were not that different. The increasing pH indicated that the bacteria might have the ability to metabolize amino acids and proteins in the medium and produce NH_3 which had basic property.

4. Conclusion

The molasses employed in this study could be used by *Nocardia* sp strain V1 as a carbon source in the fermentation medium to produce extracellular liquid having antimicrobial activity. The pre-treatment through the addition of $K_4Fe(CN)_6.3H2O$ was proven effective to remove inhibitory substances from molasses that could confine the production of antimicrobial metabolites. The bacteria grown in the synthetic, M2, and M3 media were able to produce extracellular liquid that could inhibit the growth of

Rhodotorulla sp and *Fusarium graminearum* with inhibition diameter ranges from 10-20 mm. The total sugar concentration in each medium decreased as fermentation time increased indicating that the bacteria grown in all media utilized the carbon sources inside the medium. The highest total sugar consumption rate of 96.69% was shown by the bacteria grown in the synthetic medium followed by the media containing pre-treated molasses starch and pre-treated molasses-glucose. The pH measurement showed an increasing trend during fermentation time with the highest pH of 7.58 belonged to the culture in the synthetic medium. To the best of our knowledge, this is the first report of the extracellular liquid production by *Nocardia* sp strain V1 having antimicrobial activity obtained from molasses fermentation. Further research should be focused on the biomass study and optimization process in terms of the pH and temperature to yield optimum antimicrobial activity.

References

[1] Yadav R L and Solomon S 2006 *Sugar Tech.* **8** 104-11
[2] Jiménez A M, Borja R and Martín A 2004 *Biochem. Eng. J.* **18** 121-32
[3] Satyawali Y and Balakrishnan M 2008 *J. Env. Manag.* **86** 481-97
[4] Kumar V and Satyanarayana T 2014 *Bioproc. Biosyst. Eng.* **37** 1043-53
[5] Rivas B, Moldes A B, Domínguez J M and Parajó J C 2004 *Int. J. Food Microbiol.* **97** 93-8
[6] Chao Y, Sugano Y and Shoda M 2001 *Appl. Microbiol. Biotechnol.* **55** 673-9
[7] Oikawa T, Ohtori T and Ameyama M 1995 *Biosci. Biotechnol. Biochem.* **59** 331-2
[8] Son H J, Kim H G, Kim K K, Kim H S, Kim Y G and Lee S J 2003 *Bioresource Technol.* **86** 215-9
[9] Zhu L Y, Zong M H and Wu H 2008 *Bioresource Technol.* **99** 7881-5
[10] Kotzamanidis C H, Roukas T and Skaracis G 2002 *World J. Microbiol. Biotechnol.* **18** 441-8
[11] Beaulieu M, Beaulieu Y, Melinard J, Pandian S and Goulet J 1995 *Appl. Environ. Microbiol.* **61** 165-9
[12] Sheoran A, Yadav B S, Nigam P and Singh D 1998 *J. Basic Microbiol.* **38** 123-8
[13] Lazaridou A, Roukas T, Biliaderis C G and Vaikousi H 2002 *Enzyme Microbial Technol.* **31** 122-32
[14] Bae S and Shoda M 2004 *Biotechnol. Prog.* **20** 1366-71
[15] Kalogiannis S, Iakovidou G, Liakopoulou-Kyriakides M, Kyriakidis D A and Skaracis G N 2003 *Proc. Biochem.* **39** 249-56
[16] Li W, Leet J E, Ax H A, Gustavson D R, Brown D M, Turner L, Brown K, Clark J, Yang H, Fung-Tomc J and Lam K S 2003 *J. Antibiotics* **56** 226-31
[17] Wyche T P, Hou Y, Vazquez-Rivera E, Braun D and Bugni T S 2012 *J. Nat. Prod.* **75** 735-40
[18] Nwoye C I, Agu P C, Chukwudi B C, Nwakpa S O, Ijomah I A and Idenyi N E 2015 *Biomed. Sci. Eng.* **3** 15-22
[19] Nelson N 1944 *J. Biol. Chem.* **153** 375-80
[20] Basu S, Bose C, Ojha N, Das N, Das J, Pal M and Khurana S 2015 *Bioinformation* **11** 182–4
[21] Ceitintas R and Abed Salih B 2016 *Egypt. J. Biol. Pest Co.* 26 707-12
[22] Nel B, Steinberg C, Labuschagne N and Viljoen A 2006 *Plant Pathol.* **55** 217-23
[23] Yang F, Jacobsen S, Jørgensen H J, Collinge D B, Svensson B and Finnie C 2013 *Front. Plant Sci.* **4** https://doi.org/10.3389/fpls.2013.00037
[24] Bushnell W R, Hazen B E and Pritsch C 2003 *Fusarium Head Blight of Wheat and Barley* ed K J Leonard and W R Bushnell (St. Paul, MN: APS Press) pp 44–83
[25] Pestka J J and Smolinski A T 2005 *J. Toxicol. Env. Health B* **8** 39-69
[26] Zinedine A, Soriano J M, Molto J C and Manes J 2007 *Food Chem. Toxicol.* **45** 1-18
[27] Andayani D G S, Mawarda, P C, Endah S E, Budiwati T A, Ratnaningrum D and Praharti S P 2017 *IOSR J. Agric. Vet. Sci.* **10** 7-12
[28] Perrone G, Susca A, Cozzi G, Ehrlich K, Varga J, Frisvad J C, Meijer M, Noonim P, Mahakarnchanakul W and Samson R A 2007 *Stud. Mycol.* **59** 53-66
[29] Varga J, Juhász Á and Kevei F, Kozakiewicz Z 2004 *Eur. J. Plant Pathol.* **110** 627-40

[30] Bautista-Baños S, Garcıa-Domınguez E, Barrera-Necha L L, Reyes-Chilpa R and Wilson C L 2003 *Postharvest Biol. Technol.* **29** 81-92
[31] Andersen B, Smedsgaard J and Frisvad J C 2004 *J. Agric. Food Chem.* **52** 2421-8
[32] Frisvad J C and Thrane U 2000 *Introduction to Food- and Airborne Fungi* ed R A Samson *et al* (Utrecht: Centraalbureau voor Schimmelcultures) pp 321-31
[33] Jensen B, Knudsen I M, Andersen B, Nielsen K F, Thrane U, Jensen D F and Larsen J 2013 *Int. J. Food Microbiol.* **160** 313-22
[34] Wirth F and Goldani L Z 2012 *Interdisc. Perspec. Infect. Dis.* **2012** 465717
[35] Tomsikova A 2002 *Epidemiol. Mikrobiol. Imunol.* **51** 78-81
[36] Bae S and Shoda M 2004 *Biotechnol. Prog.* **20** 1366-71
[37] Kundu S, Panda T, Majumdar S K, Guha B and Bandyopadhyay K K 1984 *Biotechnol. Bioeng.* **26** 1114-21
[38] Ashraf S, Ali S and Ulhaq I 2015 *Int. J. Res. Pharm. Biosci.* **2** 34-40
[39] Sguarezi C, Longo C, Ceni G, Boni G, Silva M F, Di Luccio M, Mazuttic M A, Maugeri F, Rodrigues M I and Treichel H 2009 *Food Bioproc. Technol.* **2** 409-14
[40] Roukas T 1998 *Proc. Biochem.* **33** 805-10
[41] Liu Y P, Zheng P, Sun Z H, Ni Y, Dong J J and Zhu L L 2008 *Bioresource Technol.* **99** 1736-42

IOP Conf. Series: Earth and Environmental Science **160** (2018) 012012 doi:10.1088/1755-1315/160/1/012012

Evaluation of hospital infectious waste management in Bandung region

N Fitria[1,2*], E Damanhuri[1] and I R S Salami[1]

[1] Faculty of Civil and Environment Engineering Bandung Institute of Technology, Jl. Ganesha 10, Bandung 40132, Indonesia
[2] Bakti Asih Higher Education School of Analyst Bandung, Jl. Padasuka Atas 233, Bandung 40192, Indonesia

*Email : novie.fitria@gmail.com

Abstract. Hospital infectious waste (HIW) has become an important issue as it possesses potential health risk and damage to the environment. HIW, if not handled in a proper way, could be a potential source of diseases such as HIV, hepatitis B&C, and other bacterial diseases that cause serious threat to human health. Therefore, an extra attention is needed for its safe and proper disposal. The sustainable management of infectious waste requires a holistic approach to monitor and evaluate the implementation of hospital infectious waste handling. Based upon sample assessment, a brief overview of current implementation of hospital infectious waste management (HIWM) in Bandung region is provided. The assessment included a 50-point check list referring to the standard Environmental Protection Agency (EPA) Guideline. The assessment of HIW management was conducted in seven hospitals. In general hospitals in Bandung region conform 57 from 100% assessment score. This is a proof that HIWM was still conducted improperly in Bandung region. Private hospitals provide better score (68%) compared to government-owned hospitals (41%) whereas accredited hospitals provide a higher score (61%) compared to hospitals with non-accreditation (22%). In addition, specific hospitals provide better compliance performance (74%) compared to general hospitals (50%). Hospitals with health worker status provided a better score (68%) than hospitals with non-outsourcing health worker status (30%). Kruskal-Wallis test showed that the value of Asymp.sig 0.000 < 0.05, significant differences were found for evaluation of HIWM between 7 hospitals. In Bandung region, the HIW management was still conducted improperly; this study provides baseline data for further research into risk assessment of HIW management.

1. Introduction

The handling of infectious waste in developing countries has many challenges; the lack of awareness of health workers and the consideration of financial budget issues become the main problem in healthcare facilities, including hospitals. Hospital is the largest producer of infectious waste when compared to other types of health facilities. Birpinar *et al* [1] and WHO [2] concluded that hospital infectious waste can be generated at up to 2.2 tons/day. Over the last decade, safe handling of infectious waste constitutes a major problem of the health sector internationally. Based on Caniato *et al* [3], healthcare waste management including infectious waste management practices varies greatly from country to country according to socio-economic conditions, regulation, and level of education. According to Lega *et al* [4] positive correlations have been found between clinical and economic performance. Recent studies highlighted the importance of management in healthcare sector. The problem is further escalated by the occurrence of new infectious diseases and the lack of appropriate infrastructure for the safe management of infectious waste. Thereby, comprising a public health risk and often resulting in environmental degradation through the release of pollutants due to the unsafe waste management practices. Based on Bdour *et al* [5], 26% of infectious waste consists of sharp and pathological wastes. In developing countries mismanagement of infectious waste is still conducted. There is no specific regulation or guidelines for segregation or classification of infectious waste [6]. This present study on infectious waste handling from seven hospitals was conducted in order to quantify the percentage of conformity with the EPA standard.

In developing countries such as Indonesia, sometimes budget issue is the reason for the misconduct of infectious waste handling. According to Government Regulation No. 44/2009, based on services, hospitals are divided into general hospital and specific hospital, meanwhile based on management, hospitals are divided into public hospital and private hospital. General hospital is a hospital in which patients with many different types of ailments are given care, or a military hospital usually located in a communication zone that gives treatment to all kinds of cases. Specific hospital is a hospital in which patients with specific type of ailments are given care, for example the hospital that specialized in eye care. Public hospital is a hospital with government ownership and private hospital is a hospital with non-government ownership. Hospital accreditation has been defined as "A self-assessment and external peer assessment process used by health care organizations to accurately assess their level of performance in relation to established standards and to implement ways to continuously improve".

Very few studies have been conducted on infectious waste in Indonesia, especially in Bandung region, which related to the assessment of HIW handling. The objectives of this study were directed towards: (1) conducting a survey of available procedures, techniques and methods of HIW handling; and (2) identifying the important characteristics and their possible influences of misconduct in the handling of HIW. The datas of current study were collected from 7 hospitals in Bandung.

2. Methods

2.1. Setting of hospital establishment

Table 1 summarizes the composition of the selected hospitals. The hospitals were grouped into two categories based on the nature of their services, namely 5 general hospitals and 2 specific hospitals.

Table 1. Group of hospital services.

Hospital Code	Hospital Services	Type/Number of beds	Accreditation level	Owner / Management	Cleaning Services Status
A	General hospital	B/192	16/Excellent	Public	Non-outsourcing
B	General hospital	B/441	16/Excellent	Public	Outsourcing
C	General hospital	C/70	5/Basic	Private	Outsourcing
D	General hospital	D/22	N.A	Public	Non-outsourcing
E	Specific hospital	C/40	5/Basic	Private	Outsourcing
F	General hospital	C/178	12/Basic	Private	Outsourcing
G	Specific hospital	A/104	16/Excellent	Public	Outsourcing

2.2. Survey method and data analysis

Check list form was made based on EPA and PP No. 101/2014 regulations to collect information on infectious waste management practices. It consists of 50 standard conditions including the segregation, collection, cleaning, transportation, and storage. We observed the selected hospitals and documentation from these facilities were reviewed and calculated based on 4 categories: a. Type of hospital based on services covered, b. Type of hospital based on management ownership, c. Type of hospital based on accreditation status, d. Type of cleaning service worker status as of outsourcing or non-outsourcing. Data was analyzed using descriptive statistic and Kruskal-Wallis method.

3. Result and discussion

3.1. Evaluation of infectious waste management

The observation for infectious waste handling was segregation, handling, containment, labeling, storage, monitoring and record keeping, training and vaccine program for waste handler, contingency plan, and personal protective equipment. Fifty items were observed closely. Check list form provided the information for its compliance with EPA standard.

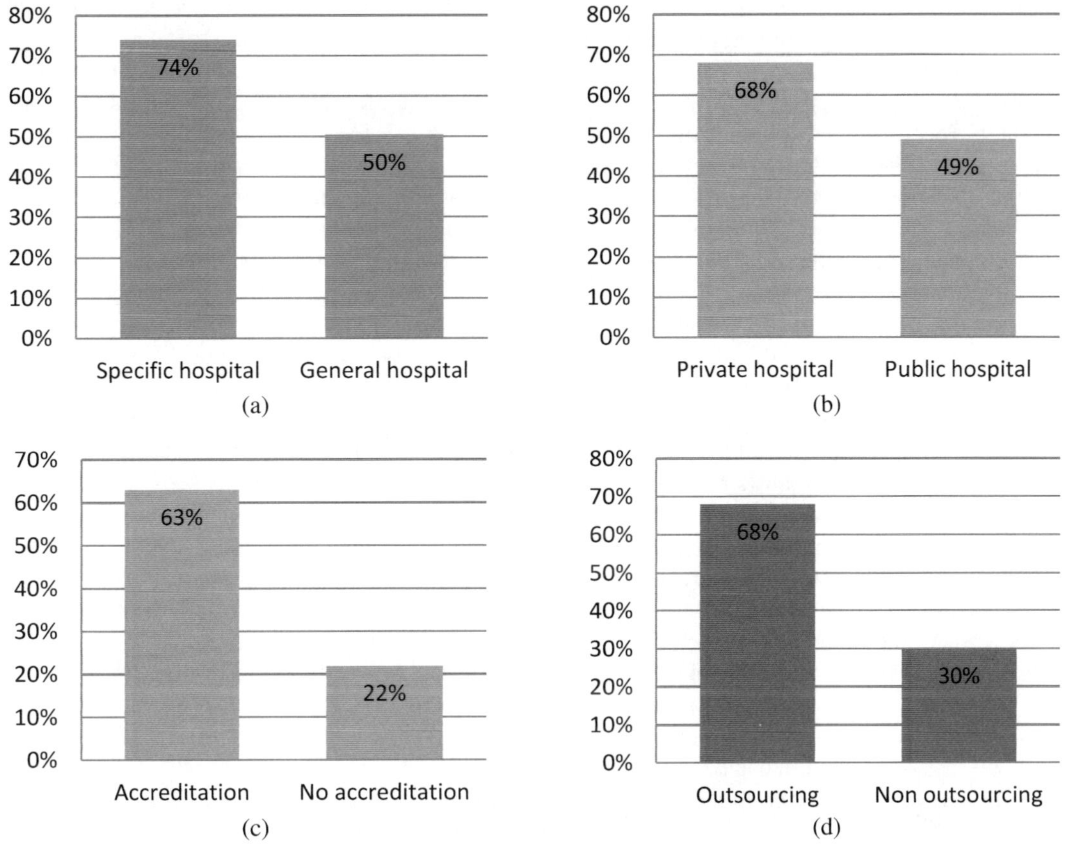

Figure 1. HIWM evaluation based on (a) hospital services, (b) owner, (c) accreditation level status, and (d) status of health worker.

Score was calculated by dividing the complied items with the total items observation. The results showed that specific hospitals had a better HIWM than general hospitals (Figure 1a), while private hospitals had higher percentage value than public hospital (Figure 1b). Hospitals with accreditation status had a better HIWM than the hospital with no accreditation (Figure 1c). Outsourcing health worker hospitals showed higher value than non-outsourcing health workers hospitals (Figure 1d). Based on 50 items of EPA standard of HIMW, general hospitals in Bandung region only met 57 of 100% assessment score. According to Indonesian law, implementation of infectious waste management in hospital was mandatory, but the implementation in real practices was not complying the regulation. Lack of knowledge on the pertinent legislation by health workers had contributed to non-conformity instances [7]. The lack of attention and sensitivity to infectious waste segregation, transportation and disposal become barriers to a proper management. The health workers were aware of the importance of segregation and other processes. However, they were careless and mixed infectious waste medical waste with domestic waste [8].

Specific hospitals had specific services for a particular type of disease. The resulting infectious waste also tended to be more homogeneous than general hospitals. In public hospitals, the type of services to the diseases varied and caused the infectious waste varied greatly. The homogeneity of waste generation is easy to manage. Therefore, based on the EPA standard for infectious waste management, specific hospital score was higher than that of the general hospital. The results also showed that private hospitals managed their infectious waste better than public hospitals. The private hospitals supervised the proper handling of infectious waste better than public hospitals with government ownership. Based on interview, the handling infectious waste was regulated by Indonesian Ministry of Health. In this case, the public hospitals were under the management of Ministry of Health Indonesia. Therefore, the public hospitals less awareness than the private hospital in handling their infectious waste. The results showed that the hospital's accreditation status affected the hospital's performance in managing its infectious waste. The accredited-hospitals require compliance with the operational standard for infectious waste management procedures. Hospitals with outsourcing cleaning service status managed their infectious waste better than hospital with non-outsourcing. The outsourcing cleaning services workers showed a higher level of awareness and cautions of infectious waste hazard compared to non-outsourced cleaning service workers.

The study revealed a general lack of knowledge of the national guidelines as well as none of method available for waste management evaluation and monitoring. Most participants believed that the waste collection procedures used in the hospital were inappropriate. A study conducted by Caniato *et al* [9] revealed that mismanagement of infectious waste management was caused by the different priorities taken by the stakeholders. However, generally segregation at the source was identified as a key requirement [9]. According to Chen and Tsai [10], biomedical waste that requires sterilization should be packed into yellow bags and infectious biomedical waste should be packed into red bags.

Based on Indonesian regulation, all types of infectious waste should be packed in yellow bags whereas red bags are for radioactive waste. The storage place should be cleaned periodically with disinfectant such as chlorine. Weighing the infectious wastes is a must. However, several hospitals in Bandung did not follow these rules. Infectious waste can be stored for one day at higher than 5°C for up to 7 - 30 days, at 0–5°C, or below 0°C, respectively. In fact, all hospitals did not have cooler rooms for infectious wastes. The infectious wastes were kept for more than 48 hr in the storage.

According to Akpieyi *et al* [11], management of infectious waste such as handling, treatment and disposal processes, may cause physical, chemical, and microbiological risks significantly. The characteristics of infection potential were based on the principles of disease transmission. However, the process of disease transmission can be conceptualized as a series of six links, with each link representing an essential step in the transfer of an infectious agent from one susceptible host to the next: (1) the presence of a stuffiest quantity of infectious agent, (2) the existence of a favorable environment reservoir for survival infectious agent, (3) a mode of escape for infectious agents, (4) a mode of transmission, (5) an infectious route entry, and (6) a susceptible host. The agency for toxic substances and disease registry listed four main transmission modes of infection, namely direct transmission, airborne transmission, vehicle airborne transmission and vector borne transmission [12].

3.2. Kruskal-Wallis test

The aim of Kruskal-Wallis test was to count significant level of infectious waste evaluation. Table 2 shows the mean rank for each hospital; smaller mean rank value indicates that the hospital conducted more improper HIWM according to EPA standard. Based on Table 2, Hospital D had the smallest mean rank value and Hospital C had the biggest mean rank value. Kruskal-Wallis test (Table 3) provides information that Asymp.sig 0.000 < 0.05, significant differences were found for HIWM evaluation.

For further analysis for Kruskal-Wallis test with value of Asymp.sig 0.000 < 0.05, Mann U Whitney analysis was conducted. The aim of Mann U Whitney analysis was to count and compare the infectious waste management evaluation for each hospital. The result of Mann U Whitney analysis is shown in Table 4.

IOP Conf. Series: Earth and Environmental Science **160** (2018) 012012 doi:10.1088/1755-1315/160/1/012012

Table 2. Mean rank value.

	Hospital	N	Mean Rank
Result	A	50	142.00
	B	50	187.50
	C	50	215.50
	D	50	114.00
	E	50	208.50
	F	50	159.50
	G	50	201.50
	Total	350	

Table 3. Kruskal-Wallis test.

		Result
Chi-Square		57.632
Df	6	
Asymp. Sig.	0.000	

Table 4. Mann U Whitney test.

Hospital	Asymp.Sig	Information
A Vs B	0.010 < 0.05	Significantly different
A Vs C	0.000 < 0.05	Significantly different
A Vs D	0.082 > 0.05	No different
A Vs E	0.000 < 0.05	Significantly different
A Vs F	0.315 > 0.05	No different
A Vs G	0.001 < 0.05	Significantly different
B Vs C	0.076 > 0.05	No different
B Vs D	0.000 < 0.05	Significantly different
B Vs E	0.193 > 0.05	No different
B Vs F	0.109 > 0.05	No different
B Vs G	0.394 > 0.05	No different
C Vs D	0.000 < 0.05	Significantly different
C Vs E	0.631 > 0.05	No different
C Vs F	0.001 < 0.05	Significantly different
C Vs G	0.351 > 0.05	No different
D Vs E	0.000 < 0.05	Significantly different
D Vs F	0.007 < 0.05	Significantly different
D Vs G	0.000 < 0.05	Significantly different
E Vs F	0.004 < 0.05	Significantly different
E Vs G	0.650 > 0.05	No Different
F Vs G	0.015 < 0.05	Significantly different

Table 4 shows the result of Mann U Whitney test. It shows that Hospital D had an evaluation value that was not significantly different from Hospital A because Asymp.Sig value was >0.05, but significantly different from Hospitals B, C, D, E, F and G because Asymp.Sig value was <0.05. Hospital D and A had the same categories, both were general, public hospitals with cleaning service status was non-outsourcing. This explains that having same category with public hospital did not guarantee that the management of infectious waste is the same. Awareness level from cleaning service worker affected infectious waste management implementation. In other word, non-outsourced cleaning service workers showed lower awareness level than outsourced cleaning service workers. Hospitals C and D also had the same category, both were general hospital type, but had different ownership and accreditation level.

Based on deep interview and observation in the field, private hospitals provide training for the workers more often than public hospitals. Private hospitals also put the budget to increase awareness level higher than public hospitals. Outsourced cleaning service workers also followed training on infectious waste handling more often than non-outsourced cleaning service.

4. Conclusion

The study revealed that general hospitals in Bandung region only conform of 57 from 100% assessment score. This is a proof that HIWM was carried out improperly in Bandung region. Hospitals with private ownership and hired outsourcing cleaning service workers had a significantly better evaluation of their infectious waste management. Improper infectious waste handling can affect the risk of disease transmission.

Acknowledgment

We would like to thank The Ministry of Science and Technology and the Directorate of Higher Education of Indonesia for the financial support.

References

[1] Birpinar M E, Bilgili M S and Erdogan T 2009 *Waste Manag.* **29** 445-8
[2] World Health Organization 2013 Safe management of wastes from health care activities, 2nd ed. http://www.searo.who.int/srilanka/documents/safe_management_of_wastes_from_healthcare_activities.pdf?ua=1
[3] Caniato M, Tudor T and Vaccari M 2015 *J. Env. Manag.* **153** 93-107
[4] Lega F, Prenestini A and Spurgeon P 2013 *Value in Health* **16** 546-51
[5] Bdour A, Altrabsheh B, Hadadin N and Al-shareif M 2007 *Waste Manag.* **27** 746-59
[6] Patwary M A, O'Hare W T, Street G, Elahi K M, Hossain S S and Sarker M H 2009 *Waste Manag.* **29** 2392-7
[7] Moreira A A M and Gunther W M R 2013 *Waste Manag.* **33** 162-7
[8] Oroei M, Momeni M, Palenik C J, Danaei M and Askarian M 2014 *J. Inf. Pub. Health* **7** 192-8
[9] Caniato M, Tudor T and Vaccari M 2012 *Waste Manag.* **35** 255-64
[10] Chen Y C and Tsai P Y 2017 *Waste Manag. Res.* **35** 593-601
[11] Akpieyi A, Tudor L T and Dutra C 2015 *Safety Sci.* **72** 127-32
[12] US Environmental Protection Agency 2016 *Model Guidelines for State Medical Waste Management* https://www.epa.gov/sites/production/files/2016-02/documents/model_guidelines_for_state_medical_waste_management.pdf

IOP Conf. Series: Earth and Environmental Science **160** (2018) 012013 doi:10.1088/1755-1315/160/1/012013

The disappearance of bromuconazole fungicide in domestic waste anaerobic fermentation

H R Hariyadi

Research Unit for Clean Technology, Indonesian Institute of Sciences (LIPI)
Komplek LIPI Bandung, Jalan Sangkuriang, Gedung 50, Bandung, 40135, Indonesia

*Email: hari.rom.haryadi@lipi.go.id, hariyadi_hr@yahoo.com

Abstract. A previous study recommended observing degradation of bromuconazole in waste anaerobic fermentation. This research, therefore, was carried out to study the disappearance of bromuconazole in waste anaerobic fermentation. Bromuconazole is a fungicide which is effective against *Ascomycetes*, *Basidiomycetes* and fungi imperfecti in cereals, grapes, top fruits and vegetables. It is also effective against *Alternaria* and *Fusarium* sp. In this study, one-month old refuse added with bromuconazole was anaerobically incubated in the 500 ml bottles at 30°C with no shaking in the dark room. Bromuconazole with total concentration of 200 mg/L was added in day 0 and 31. As a control, a bottle with refuse and bromuconazole at concentration of 200 mg/L was autoclaved and incubated at the same condition. High-performance Liquid Chromatography with UV detector and a 100 RP18 5μm Lichrosphere column was used to determine bromuconazole concentration. Methane content was determined by Gas Chromatography method equipped with a flame ionization detector and a metal column packed with 5% neopentyl glycol sebacate and 1% H_3PO_4 on Chromosorb W-AW (mesh 80-100). The disappearance of bromuconazole in anaerobic enrichments can be observed from mass balance determination after incubation for 225 days. The result showed that there was no change in bromuconazole mass balance. Bromuconazole was mostly adsorbed by the refuse. Bromuconazole inhibited the production of methane (99.58 mM), less than that of the control bottle with no added bromuconazole (121.78 mM). This phenomenon indicated the persistence of bromuconazole and its toxicity against methanogen bacterial activity. The presence of bromuconazole in the anaerobic process caused the refuse to be imperfectly fermented and so the production of methane was inhibited and volatile fatty acids were accumulated. Accumulation of the volatile fatty acids may cause further pollution in aquatic environment around the landfills site.

1. Introduction

Azole compounds, particularly triazole, have been used as the active ingredient in the manufacture of fungicides and pharmaceutical preparations since the 1970s. Although the production and its use have increased, there is still a lack of information about the nature of these compounds in the environment as well as landfill site and process effluent. The problem may arise when these potentially toxic compounds pollute the environment and threaten both the aquatic organism as well as human health. Activities of microorganisms might be disturbed by the presence of such compounds, and decrease the fermentation process performance. Therefore, the experiment to observe the presence of the compound in anaerobic process of the refuse is important. There are about 16 different fungicide triazole derivatives and one of them is bromuconazole (Figure 1). This compound has been developed by the Rhone Poulenc

Agrochemie UK since 1991. Bromuconazole ($C_{13}H_{12}BrCl_2N_3O$, MW = 377) or 1[(2RS,4RS,2SR,4SR)-4-bromo-2-(2,4-dichlorophenyl)tetra-hydrofurfuryl]-1H-1,2,4-triazole is a heterocyclic-aliphatic compound belongs to the groups of systemic fungicide and it is active against *Ascomycetes*, *Basidiomycetes*, fungi imperfecti in cerealia and fruits as well as active against fungus *Alternaria* sp and *Fusarium* sp. Toxicity by 50% (LD_{50}) for rats and mice are 365 mg/kg and 1151 mg/kg, respectively. Towards fish such as Bluegills and trout, the toxicity of bromuconazole (LC_{50}) are 3.1 mg/L and 1.7 mg/L, respectively [1]. Bromuconazole is one of the 79 substances of the third stage Part A of the review program covered by Commission Regulation (EC) No 1490/2002 [2]. Similar products such as Amitrole (3-amino-1 h -1,2 4-triazole) was reported to be transformed into 3-(3-amino-1 h-triazole-1-yl) alanine by *E. coli* [3]. Benzotriazole transformed to 1-deoxyribosida by *Aerobacter aerogeneses* ATCC9621 supplementation with 5'-thymidylic acid and *Aspergillus niger* (van Teigh) was able to reduce Triadimefon to become Triadimol [4].

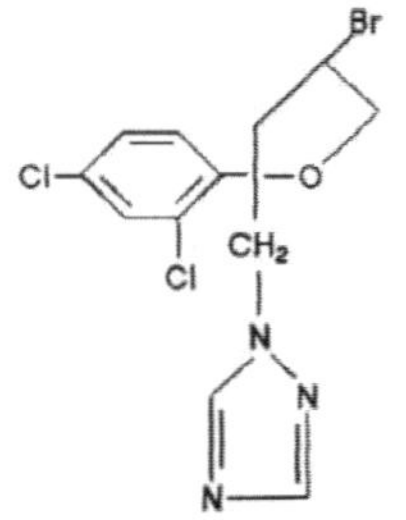

Figure 1. Chemical structure of bromuconazole.

As a pollutant, bromuconazole might be degraded biologically in aerobic and anaerobic conditions. In anaerobic fermentation of organic compounds, the resulted xenobiotic compounds such as pesticide will accumulate if not degraded into methane [5]. The production of methane as a source of energy is one of the advantages of anaerobic fermentation process. Anaerobic degradation occurs due to the role of either one type or mixture of microbes. For example, the benzoate was degraded by *Moraxella* sp or *Pseudomonas* to diethyl ester [6] and degradation of phenols by a community of microbes produced methane [7–8]. Xenobiotic compounds can be degraded if the appropriate enzyme are produced by microbes during the process of evolution. The degradation is influenced by the ability of the enzyme to accept a substrate with a structure similar but not identical to that of a natural and less complex compound that is also present in media. The degradation is also influenced by the ability of substrate to stimulate synthesis of the necessary enzyme [9]. A previous study [10] recommended to observe the degradation of bromuconazole in waste anaerobic fermentation. This research, therefore, was carried out to observe the disappearance of bromuconazole in waste anaerobic fermentation.

2. Experimental

2.1. Anaerobic experiment

This study was conducted in Waste Technology Laboratory Strathclyde University Glasgow. The presence of anaerobic bromuconazole was carried out using one month partially degraded waste substrate taken from local domestic waste disposal as source of microbes. An amount of 25 g of refuse was filled into 500 ml volume of bottle and added with 250 ml of water. The bottle was then sealed with rubber sub-seal No. 45. Prior to incubation at 30°C without stirring in a dark room, the bottle was purged with oxygen-free nitrogen gas assuming the anaerobic conditions was obtained. The experiment was conducted using three replicates. Bromuconazole is an aromatic compound with solubility in water is 50 mg/L, so for preparing a solution with a high concentration, organic solvent such as acetone had to be used. Bromuconazole dissolved in acetone with final concentration of 100 mg/L was added each on day 0 and 31 (treatment 1). A different bottle (treatment 2) on day 31 was added with 100 mg/L of phenol (final concentration). Another bottle as treatment 3 was added with 146 mg/L of benzoate as final concentration. The bottles with phenol and benzoate were then each added with 200 mg/L of bromuconazole as final concentration on day 103. A bottle added with 100 mg/l of bromuconazoleat each on day 0 and 31 was then autoclaved as treatment 4 or sterile control. Samples were taken on a regular basis to determine the concentration of bromuconazole in liquid and methane in the headspace bottle.

2.2. Chemical analysis

2.2.1. Bromuconazole. Determination of the concentration of bromuconazole was carried out by High-performance Liquid Chromatography (HPLC) method. Bromuconazole standard was received from the Rhone-Poulence Agrochemie whereas HPLC solvents such as acetonitrile and other chemicals were purchased from E-Merck. The equipment consisted of a Binary LC250 pump (Perkin Elmer), Lichrosphere 100 RP18 column um 5:150 x 4.6 mm (Altech), and UV detector LC 90J (Perkin Elmer). An interface of Nelson and Nelson 900 Series system integrators software PC version 5.1 was used to process the data. The determination was carried out isocratically with mobile phase of acetonitrile/water (65/35% v/v) on the following conditions: flow rate of 1 ml/min, wavelength UV at 230 nm, response of 0.5, sensitivity of 1 AUFS, maximum pressure of 3500 psi, duration of 15 min, and injection volume of 20 μl. Bromuconazole concentration in the sample was determined based on the area and the concentration of the standard of bromuconazole dissolved in mobile phase on the same analysis conditions.

2.2.2. Methane. The concentration of methane in the headspace bottle was determined using Gas Chromatography method, GC8700 (Perkin Elmer) equipped with Flame Ionization Detector (FID). The temperature of the injector, detector, and oven was each set on 200, 210, and 80°C, respectively. Pure methane (ul 10 x 5) and the triplicates sample was separately injected through a gas syringe (Hamilton). Concentration of methane was calculated based on the standard temperature (K) and local pressure (mm). Oxygen-free nitrogen gas was used as the carrier with speed 20 lbf/sq-inc, column (Phase Sep) made of metal of 2 m x 2 mm. The type of packing column used was 5% neopentyl glycol sebacate and 1% H_3PO_4 on Chromosorb w-aw (80 – 100 mesh).

3. Results and discussion

By HPLC, bromuconazole was separated into two peaks with a proportion of 51:49% at 5.07 and 6.58 min, respectively. Bromuconazole was present in four absolute compounds that comprised two sets of diastereomers and two sets of enantiomers. HPLC separated the 2 sets of diastereomer but not the 2 sets of enantiomers. These two peaks, corresponded to diastereomer 1, called as bromconazole 1, consisted of 1,2,4- and 1,3,4-bromuconazole and diastereomer 2, called as bromuconazole 2, consisted of 1,2,3- and 1,2,5-bromuconazole. Bromuconazole is a lipophilic compound with the solubility in water as low as 50 mg/l and may attach to the solid particle easily. Therefore, the disapperance of bromuconazole domestic waste anaerobic fermentation with refuse as inoculum source could be observed from mass balance determination. The mass balance of bromuconazole was determined after 225 days (An6) and 122 days (An7 and An13) of incubation. The result shows that there were no changes in the total bromuconazole mass balance (Figure 2).

Bromuconazole 1 (Figure 3) and bromuconazole 2 (Figure 4) in all enrichments and sterile control were, mostly, adsorbed by refuse particle. No net disappearance of bromuconazole, therefore, was observed in refuse anaerobic fermentation. This result suggested that bromuconazole was not degraded by microbes present in the refuse. Although one month partially degraded refuse was used as the source of organic carbon and microbes, this refuse contained cellulosic compounds did not support the capabilities of microbes to consume the bromuconazole. Based on this result, it was suggested that the properties of bromuconazole, such as lipophilicity, determine the effective treatment of wastes containing this compound. The presence of solid particles or biomass might remove bromuconazole by adsorption. However, the adsorption of bromuconazole by refuse might affect the production of methane in anaerobic fermentation.

IOP Conf. Series: Earth and Environmental Science **160** (2018) 012013 doi:10.1088/1755-1315/160/1/012013

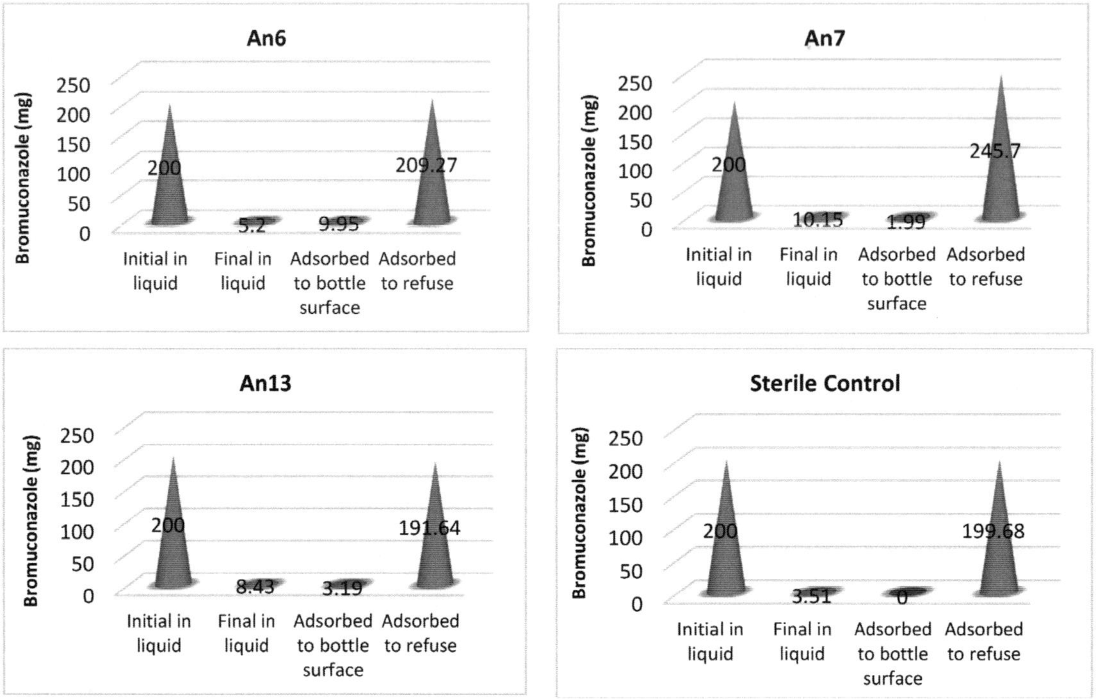

Figure 2. Mass balance of bromuconazole in domestic waste anaerobic fermentation.
An6: Bromuconazole (100mg/L) was added on day 0 and day 31.
An7: Bromuconazole (200mg/L) was added on day 103 in media supplemented with phenol (100mg/L) on day 31.
An13: Bromconazole (200 gm/L) was added on day 103 in media supplemented with benzoate (146 mg/L) on day 31.

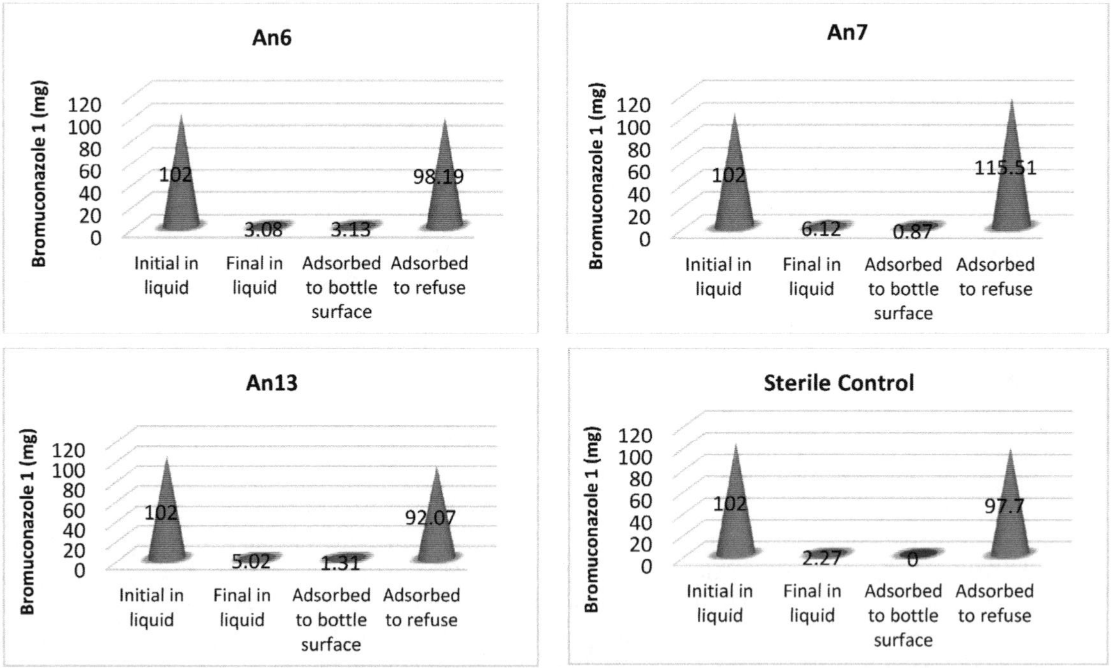

Figure 3. Mass balance of bromuconazole 1 in domestic waste anaerobic fermentation.
An6: Bromuconazole (51mg/l) was added on day 0 and day 31.
An7: Bromuconazole (102mg/l) was added on day 103 in media supplemented with phenol (100mg/l) on day 31.
An13: Bromconazole (102gm/l) was added on day 103 in media supplemented with benzoate (146 mg/l) on day 31.

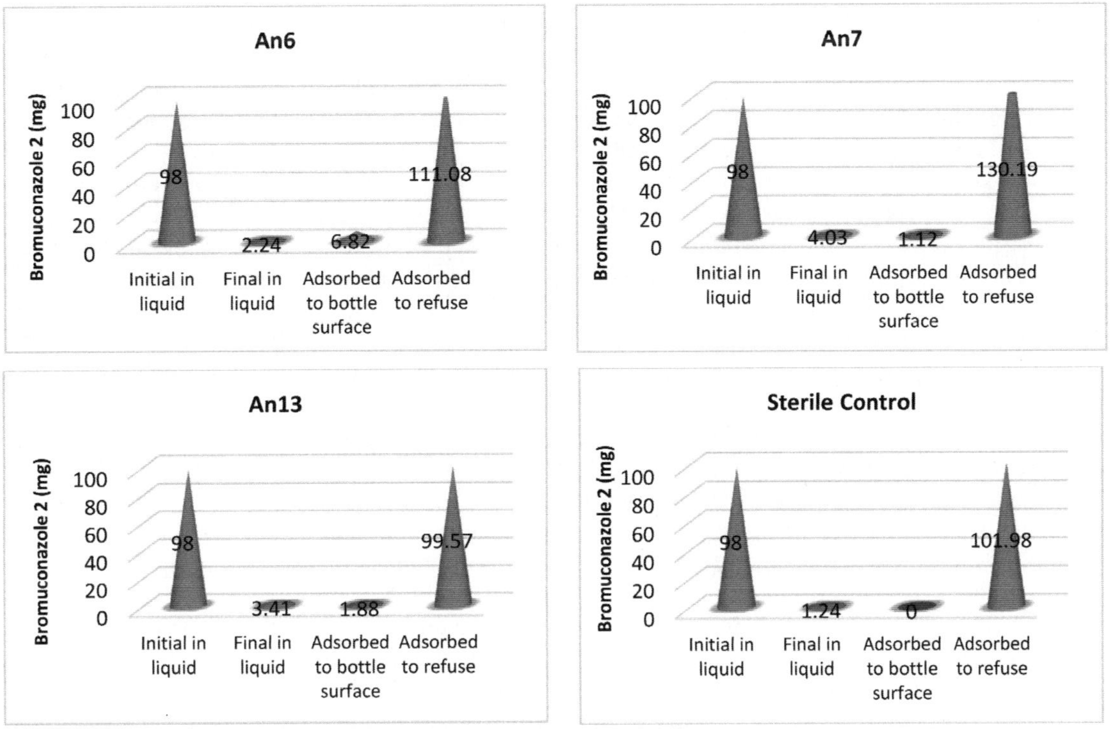

Figure 4. Mass balance of bromuconazole 2 in domestic waste anaerobic fermentation.
An6: Bromuconazole (49mg/l) was added on day 0 and day 31.
An7: Bromuconazole (98mg/l) was added on day 103 in media supplemented with phenol (100mg/l) on day 31.
An13: Bromconazole (98gm/l) was added on day 103 in media supplemented with benzoate (146 mg/l) on day 31.

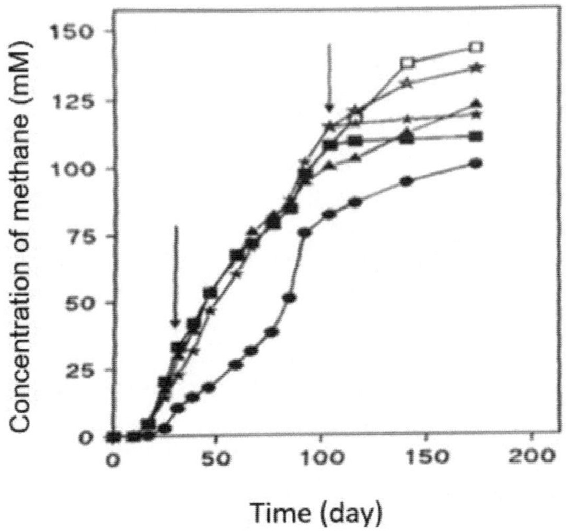

Figure 5. Methane production from refuse supplemented with: ● bromuconazole 100 mg/L at day 0 and 31; ■ phenol 100 mg/L at day 31 and bromuconazole 200 mg/L at day 103; ★benzoate 146 mg/L at day 31 and bromuconazole 200 mg/L at day 103; □☆ phenol or benzoate at day 31 and no added bromuconazole at day 103; ▲ control: without phenol, benzoate or bromuconazole.

In this experiment, benzoate and phenol had been selected as the additional carbon source or induced substrate due to the basic structure of the compounds was similar with one (phenyl ring) of the rings that formed bromuconazole. It was expected that with the additional carbon source depletion, microbes degraded bromuconazole through its phenyl ring. Figure 5 shows the effect of bromuconazole addition on the methane production in refuse fermentation.

In this experiment, bromuconazole was added on day 0 and 31 so that the total concentration in the bottles as final concentrations was 200 mg/L. The methane production after bromuconazole addition on day 0 was lower than the control. At day 31, another 100 mg/L of bromuconazole was added, methane was continuously produced. Although methane was produced continuously but on the day 173 its concentration was 99.58 mM, lower than the control. It suggested that bromuconazole had a greater influence on methanogen activities. As shown in Figure 5, methane production stopped as so bromuconazole 200 mg/L as inal concentration was added on the day 103. Inhibition of methane production continued until the end of the experiment at day 173. Methane was still generated on bottles without bromuconazole addition on day 103. Presumably, the bromuconazole attached to the surface of refuse particles protected the refuse from being degraded by the microbes to obtain energy. Therefore, high concentrations of co-substrates (bromuconazole) could starve the methanotrophs of the energy that was needed to survive [11]. This experiment indicated that the contamination of bromuconazole was highly influential on the production of methane and the concentration of 200 mg/L might be too toxic to the methanogen bacteria. Bromuconazole is a lipophilic compound with rings of benzyl, triazole, and phenyl. According to a study, there was a significant correlation between the toxicity of phenyl compounds and their hydrophobicity, but no correlation was found for benzyl compounds. This is the reason that the more hydrophobic or lipophilic a molecule is, the more readily it crosses the cell membrane and becomes highly toxic and vice versa [12]. Anaerobic fermentation showed ineffective degradation of bromuconazole and therefore, it is interesting to conduct the experiments on aerobic treatment.

4. Conclusion

It was concluded that phenol or benzoate as a source of additional carbon improved methane production. By improving methane production, phenol or benzoate indicated no inhibition on methanogen bacterial activity. The presence of phenol or benzoate in refuse was tolerated by fermentative bacteria as well as methanogen bacteria to degrade refuse. In contrast, bromuconazole indicated its toxicity against methanogen bacterial activity. The addition of phenol and benzoate improved adaptation of the anaerobic microbes to bromuconazole. The presence of bromuconazole in anaerobic process caused the refuse imperfectly fermented and so the production of methane was inhibited and volatile fatty acids were accumulated. Accumulation of the volatile fatty acids may cause further pollution in aquatic environment around the landfill site. This phenomenon indicated the persistence of bromuconazole and this compound was not degraded in the domestic waste anaerobic fermentation process.

References

[1] Worthing C R 1991 *The Pesticide Manual: A World Compendium 9th ed.* (Suffolk: British Crop Protection Council)
[2] European Food Safety Authority 2010 *J. EFSA* **8** 1704–88
[3] Kieslich K 1976 *Microbial Transformation of Non-Steroid Cyclic Compounds* (New York & London: John Wiley & Sons)
[4] Deas A H B and Clifford D R 1982 *Pest. Biochem. Physio.* **17** 120–33
[5] Fewson C A 1981 *FEMS Symp.* **12** 141–79
[6] Evans W C and Fuchs G 1988 *Ann. Review Microbiol.* **42** 289–317
[7] Balba M T and Evans W C 1977 *Biochem. Soc. T.* **5** 302–4
[8] Fery J G and Wolfe R S 1976 *Arch. Microbiol.* **107** 33–40
[9] Hutzinger O and Veerkamp W 1981 *FEMS Symp.* **12** 3–45
[10] Hariyadi H R 2017 *IOP Conf. Ser.: Earth Environ. Sci.* **60** 012022

[11] Strong P J, Clarke W, Karthikeyan O P, Zhu J and Wu W 2017 *Agro-Environmental Sustainability* ed J S Singh and G Seneviratne (Cham: Springer) pp 19-40
[12] Kayembe1 K, Basosila L, Mpiana P T, Sikulisimwa P C and Mbuyu K 2013 *Int. Res. J. Pure Appl. Chem.* **3** 48-58

IOP Conf. Series: Earth and Environmental Science **160** (2018) 012014 doi:10.1088/1755-1315/160/1/012014

Influence of pH on biogas production in a batch anaerobic process of tofu wastewater

Widyarani[1,*], Y Victor[2], L Sriwuryandari[1], E A Priantoro[1], T Sembiring and N Sintawardani[1,*]

[1] Research Unit for Clean Technology, Indonesian Institute of Sciences (LIPI), Komplek LIPI Bandung, Jalan Sangkuriang, Gedung 50, Bandung 40135, Indonesia
[2] Department of Physics-Energy Engineering, Surya University, Jl. Boulevard Gading Serpong Kav. M5 No. 21, Tangerang 15810, Indonesia

E-mail: widyarani@lipi.go.id, neni001@lipi.go.id

Abstract. As a popular soybean-based food in Indonesia, tofu production might cause a potential environmental problem, especially to the water body around the tofu industries. Approximately 17-45 tonnes of tofu wastewater are generated for every tonne of soybean used. The wastewater has a high organic concentration at pH of 3.5-4.5. Anaerobic digestion not only could be an efficient method to treat wastewater with high organic content, but also it produces less sludge and generate biogas as a renewable energy source. Anaerobic microorganisms normally work well in neutral pH. By pH lower than 5 the metabolism of methanogenic community could be easily hampered. The objective of this research was to test the influence of pH on anaerobic digestion of tofu wastewater. The experiments were performed in a batch mode using tofu whey waste water indigenous at pH 3.7 and neutralized wastewater at pH 7 Concentrations were varied at 20, 50, and 100% wastewater. Our results showed that at low pH of tofu wastewater did not have a negative influence on anaerobic digestion of tofu wastewater in a batch mode. A further addition of wastewater into the reactor (batch-wise) also showed a similar trend. It suggests that the presence of adequate buffer capacity was more important than adjustment of wastewater pH.

1. Introduction

Tofu or soybean curd is a soybean-based food that is common and popular in South East Asia. It average consumption is of 7 kg/person/year [1]. In Indonesia, there are around 80,000 tofu factories—mostly are small and medium enterprises—with a production capacity of 2.6 million tonnes of soybean per year [2]. Tofu production requires 17-45 kg of water per kg of soybean [3,4], some of which are discharged as wastewater. Tofu wastewater contain 7.5-14 g-COD/l [4], which should be treated properly before discharged into water bodies.

Anaerobic digestion is an efficient method to treat wastewater with high organic content with more than 5 g-COD/l. The process requires low energy and nutrient input, and correspondingly low biomass or sludge production [5]. The high energy-efficiency and environmentally beneficial technology for bio-energy production are the advantages of anaerobic process [6]. Anaerobic digestion performance is influenced by several factors such as temperature, pH, ratio of carbon and nitrogen, organic loading rate, and hydraulic retention time [7].

Anaerobic digestion consists of several reactions such as hydrolysis, acidogenesis, and methanogenesis. These reactions are performed by different microbial consortia. Due to these microorganisms and their enzymatic activities, anaerobic digestion is highly dependent on pH [7]. Hydrolysis is reported to be optimum around neutral pH [8,9], although this is also subject to the type of substrate. The optimum pH of carbohydrate hydrolysis is between 6-7.5 [10,11]. Protein and lipid hydrolysis are more efficient at neutral or slightly alkaline pH [12–14]. The optimum pH of acidogenesis is 6 [15]. Methanogenesis takes place between pH 6.7-7.8; the optimum pH is 7 [14,16]. In a batch system or a single-step continuous reactor where these reactions may take place simultaneously, pH 6.8-7.4 has been suggested as optimum [7].

Tofu making process in Indonesia uses acidic solution to coagulate soymilk [3,4]. This process results in wastewater with a pH between 3-5 [4], which is lower than the optimum pH of anaerobic digestion. The objective of this research was to test the influence of the pH of the wastewater on anaerobic digestion of tofu wastewater. Experiments were performed in a batch mode using waste water indigenous at pH 3.7 and neutralized wastewater at pH 7. Biogas production was observed and compared.

2. Material and Methods

2.1. Materials

Tofu wastewater used in the experiments was obtained from a tofu factory located in Bandung, West Java, Indonesia. The wastewater used was the liquid (whey) that was discharged after coagulation of soymilk.

The inoculum used in this experiment was cow rumen obtained from a local slaughterhouse. Cow rumen was diluted with water (1 : 1), and molasse was added as nutrition source. The mixture was placed in a closed container at room temperature, and incubated for two months prior to experiments.

2.2. Experimental Setup

During the experiments, 2.5 L bottles were used as anaerobic reactors (Figure 1).

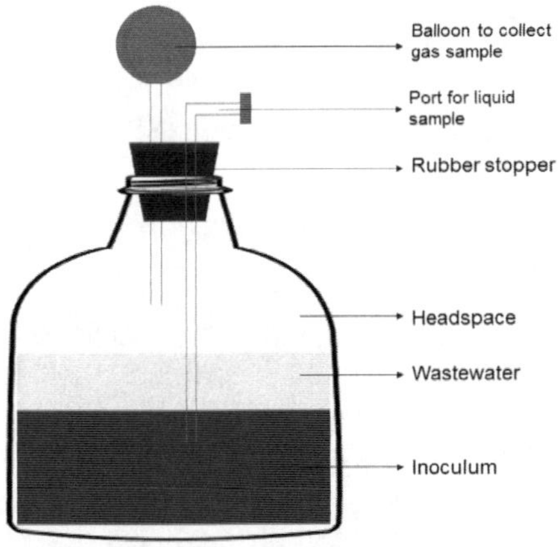

Figure 1. Experimental setup.

Table 1. Wastewater addition (ml).

	Concentration (%)					
	20		50		100	
Day	Wastewater	Water	Wastewater	Water	Wastewater	Water
0	100	400	250	250	500	0
13-16[a]	30	0	75	0	150	0

[a] Volume of wastewater addition per day

The experiments were performed using wastewater indigenous at pH 3.7 and pH 7. Solution of 5 N NaOH was used to increase the wastewater pH to 7. Waste water concentration was varied by 20, 50, and 100%.

Each of 2.5 L reactor contained 1 L of inoculum and 0.5 L mixture of wastewater and water (Table 1). An experiment with only 0.5 L water was used as control. The high amount of inoculum was to ensure that anaerobic digestion took place. Subsequently, the reactor was flushed with nitrogen gas to remove remaining oxygen, immediately sealed with rubber stopper, and incubated at room temperature. During the 13 incubation days, gas production from the balloon (Figure 1) was measured each day by using a syringe. After 13 days, wastewater with corresponding volume (Table 1) was added to the reactors. The incubation and gas monitoring were continued for 4 days. The experiments were performed in duplicates.

2.3. Analysis and calculations

Biogas produced during the experiment was collected in a balloon attached to the reactor (Figure 1). A syringe was used to collect and measure the volume of biogas in the balloon.

Liquid samples were drawn during day 13-16 for measurement of pH, bicarbonate alkalinity, and total volatile fatty acids (VFA). For alkalinity and VFA measurement, samples were filtered using paper filter. Using 0.1 N H_2SO_4, titration was performed on 20 ml of filtrate to pH 5 and pH 4.4. The volume of H_2SO_4 that was used for titration from the initial pH to pH 5 (A ml) and from pH 5 to pH 4.4 (B ml) were used in calculations as follows [17]:

$$\text{Bicarbonate alkalinity} = A \times 250 \text{ mg/ml (as } CaCO_3)$$

$$\text{Total VFA} = 500 \times (1{,}66\,B - 0{,}15) \text{ mg/l (as acetic acid)}$$

3. Results

3.1. Biogas production

Biogas formation during the first 13 days of the experiment is shown in Figure 2. Figure 2a and 2b show at 20 and 50% wastewater concentration and there was no significant difference between biogas formation from wastewater at pH 3.7 and 7.

At 100% wastewater concentration, between day-8 and day-13 more biogas was produced with wastewater pH 3.7 (Figure 2c). However, biogas formation with wastewater at pH 3.7 could not be measured, possibly due to leakage in the reactor. Therefore, the result for the first 13 days generated from a single experiment instead of duplicate. Between day 14 and 16 from duplicate results, it is apparent that biogas production from wastewater pH 3.7 was higher than pH 7.

3.2. Buffer capacity

After 13 days of experiment, wastewater was added to the reactor and pH was measured (Figure 3).

Figure 3a-c show that on day-13, the pHs for all experiment were in the range of 7.1-7.3. After wastewater addition, the lowest pH of 6.7 was observed on day-16 for 100% wastewater at pH 3.7 (Figure 3c). In this experiment, the control experiment pH also decreased to 6.4. For all experiments with wastewater pH of 7, the pH after wastewater addition ranged between 7.4-7.9.

Bicarbonate alkalinity is an indicator of alkaline buffer capacity toward changes of pH. Figure 4a shows the alkalinity of experiments with wastewater pH 3.7 decreased between day-13 and day-16. On the other hand, the alkalinity of experiments with wastewater pH 7 slightly increased between day-13 and day-16 (Figure 4b).

Figure 5a shows that after addition of wastewater (day 14-16), total VFA for experiment at wastewater pH 3.7 slightly increased. On the other hand, at wastewater pH 7 the increases of total VFA after wastewater addition were more apparent (Figure 5b).

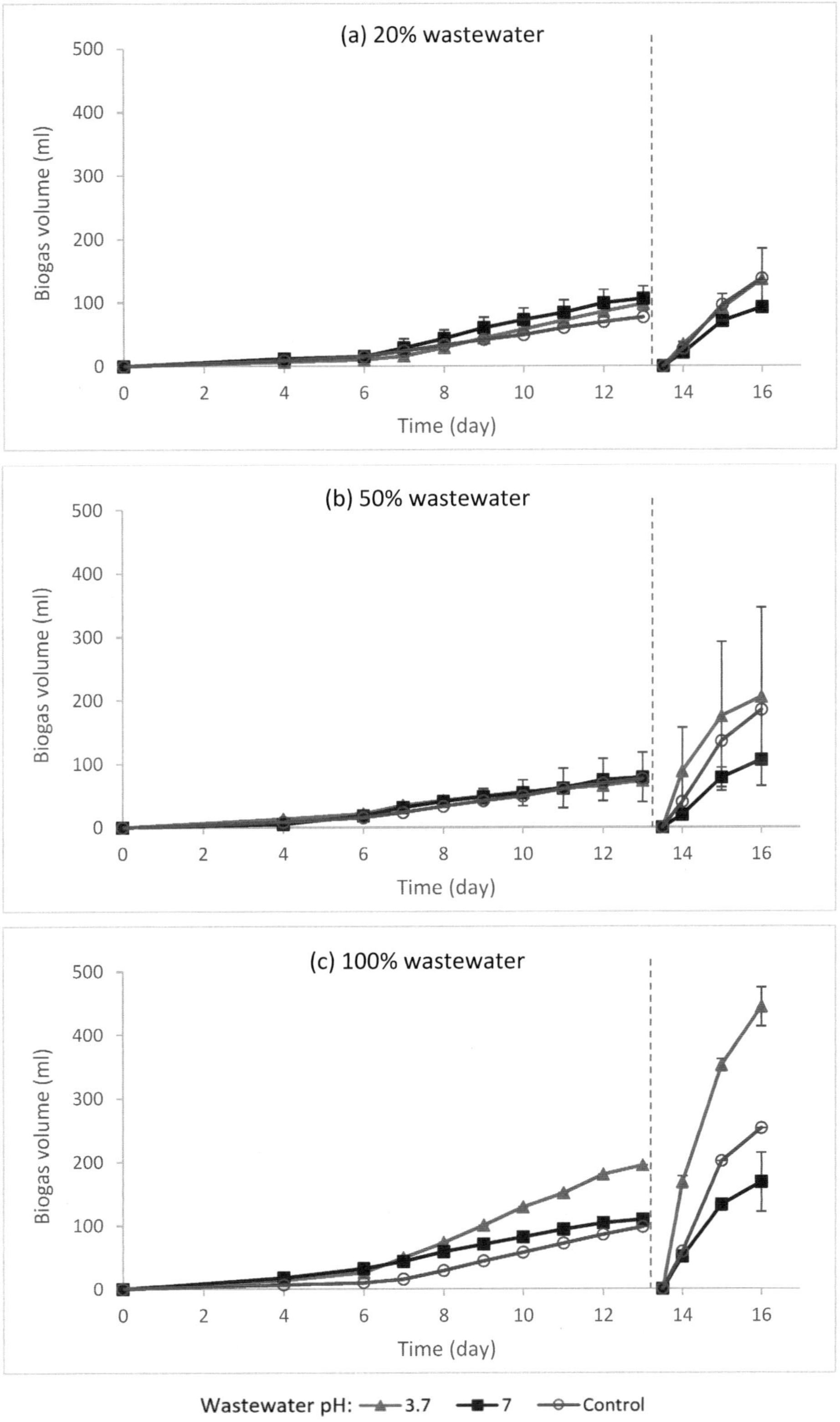

Figure 2. Cumulative biogas formation at wastewater concentration of (a) 20%, (b) 50%, and (c) 100%.

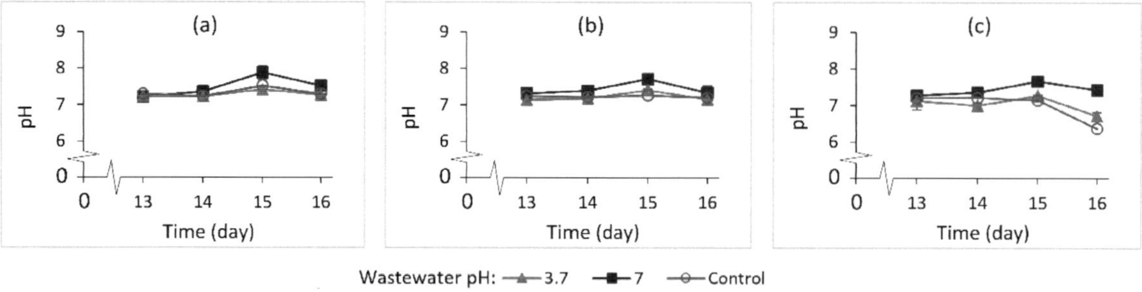

Figure 3. pH of the experiments at wastewater concentration of (a) 20%, (b) 50%, and (c) 100%.

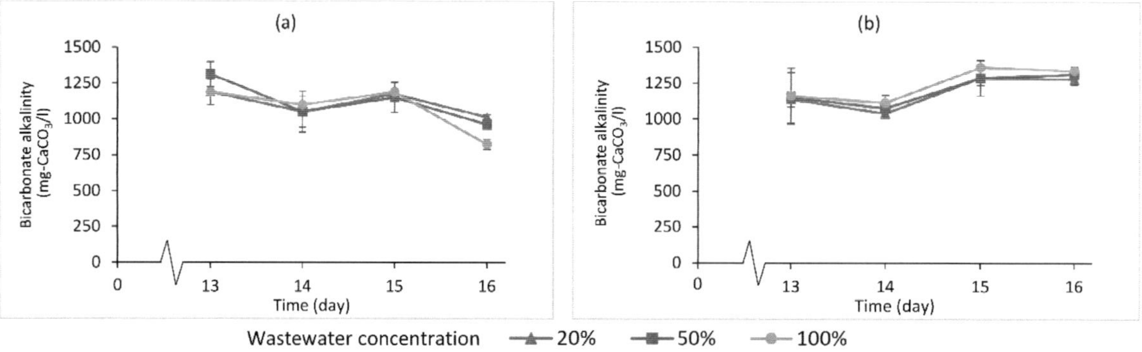

Figure 4. Bicarbonate alkalinity of the experiments at wastewater pH (a) 3.7 and (b) 7.

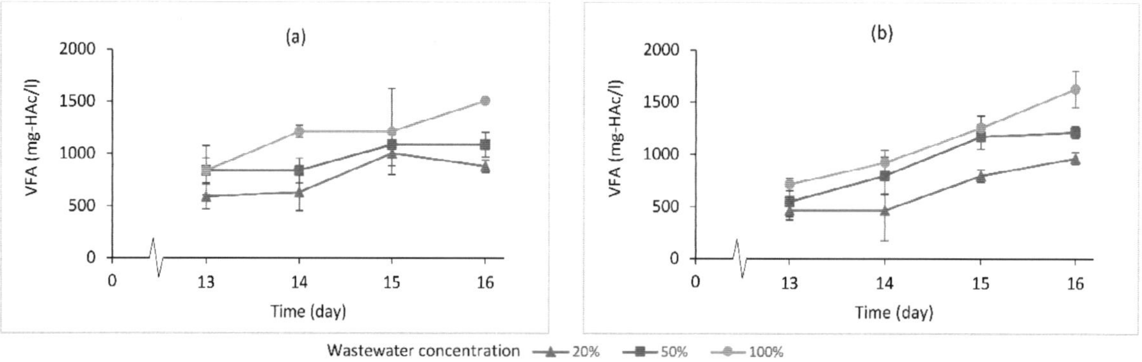

Figure 5. Total volatile fatty acids (VFA) at wastewater pH (a) 3.7 and (b) 7.

4. Discussion

Low pH is not expected in anaerobic digestion, since hydrolysis, acidogenesis, and methanogenesis processes are more optimum in pH around 7 [7]. However, Figure 2a-c show that the lower pH of the wastewater did not hamper anaerobic digestion. This might be due to the pHs of the mixture in all experiments were 7 or higher (Figure 3a-c), regardless the pH of the wastewater. This also shows that the system had sufficient buffer capacity to keep the pH stable.

Not only the low wastewater pH did not hamper anaerobic digestion, but also at 100% wastewater concentration there was indication that biogas production with wastewater pH 3.7 was higher than with wastewater pH 7. This might be due to the use of NaOH solution to adjust the pH. Sodium ions (Na^+) might have toxic effect that inhibit anaerobic processes, particularly on methanogenic microorganisms [18]. In our experiment, however, Na^+ concentration was probably too low to cause toxicity.

The other possible mechanisms is due to the accumulation of inhibiting compounds as the result of hydrolysis. NaOH aids carbohydrate hydrolysis, resulting in some sugar monomers and might also in some phenolic compounds from lignin degradation [19]. Some phenolic compounds might have inhibiting effects on anaerobic digestion [20].

Tofu wastewater in our experiment had a protein content of 1.6 g/l (unpublished result). The amount of NaOH might correlate with increased protein solubilisation and hydrolysis [21]. During anaerobic digestion of protein-rich substrate, ammonia might accumulate as a product of protein hydrolysis and acidogenesis [22]. The increase of ammonia with wastewater pH 7 is indicated by the higher pH (Figure 3). Since total ammonia (NH_3 and NH_4^+) inhibited anaerobic processes, ammonia accumulation might result in lower biogas production [22,23]. On the other hand, the presence of both Na^+ and ammonia was suggested to show less inhibition compared with when only one of the two inhibitors was present [23]. Furthermore, ammonia contributes to process stability by creating buffer at pH 6.4-9.3 [24].

In either case where the excess hydrolysis products from lignin and/or protein might cause inhibition of anaerobic digestion, the low biogas production might be temporary. After an adaptation period, the excess hydrolysis become available substrates for acidogenesis and methanogenesis processes, and increased biogas production.

Another indicator of process stability is the ratio of total VFA and alkaline buffer capacity. During the experiments, the alkalinity also changed as a result of the depletion of organic acid and volatile fatty acids. Lactic acid, valeric acid, butyric acid, propionic acid, and acetic acid in the medium were consumed by bacteria in their metabolism [14]. The value of stable operation is empirical and system-specific; however, ratio of 0.4-0.6 is often used as an indicator of stable continuous operation [17]. In this experiment, VFA/buffer capacity ratios were between 0.5 and 1.8. These values might indicate that the reactors were too low on substrate, as also indicated by the increase of biogas formation after wastewater addition, particularly at 100% wastewater concentration (Figure 2c).

5. Conclusions

Our results show that the low pH of tofu wastewater did not have a negative influence on anaerobic digestion of tofu wastewater in a batch mode. Contrary to expectation, pH adjustment of wastewater with high protein content might have negative influence, possibly due the presence of sodium ion. Our results suggest that the presence of adequate buffer capacity was more important than the adjustment of wastewater pH.

Acknowledgments

This research was funded by Indonesian Institute of Sciences (LIPI) through the Deputy of Engineering Sciences priority program and thematic program. The authors would like to thank M. Thoriq Aulie Reza and Mimin Suminar for their help during the experiments.

References

[1] Astuti S P and Hakiki G 2016 *Consumption of Calorie and Protein of Indonesia and Province* ed N Iriana and H Rangkuti (Jakarta: Badan Pusat Statistik)
[2] Indriyati and Diyono 2012 *J. Tek. Lingkungan* **Hari Bumi** 89–94
[3] Djayanti S 2015 *J. Ris. Teknol. Pencegahan Pencemaran Ind.* **6** 75–80
[4] Sriharti, Salim T and Sukirno 2004 *Prosiding Seminar Nasional Rekayasa Kimia dan Proses 2004 (Semarang)* I-12-1
[5] Van Lier J B 2008 *Water Sci. Technol.* **57** 1137–48
[6] Fehrenbach H, Giegrich J, Reinhardt G, Sayer U, Gretz M, Lanje K and Schmitz J 2008 *UBA-Forschungsbericht* **206** 41–112
[7] Mao C, Feng Y, Wang X and Ren G 2015 *Renew. Sustain. Energy Rev.* **45** 540–55
[8] Dinamarca S, Aroca G, Chamy R and Guerrero L 2003 *Water Sci. Technol.* **48** 249–54
[9] Zhang B, Zhang L-L, Zhang S-C, Shi H-Z and Cai W-M 2005 *Environ. Technol.* **26** 329–40
[10] Faulds C B, Robertson J A and Waldron K W 2008 *J. Agric. Food Chem.* **56** 7038–43
[11] Hu Z-H, Yu H-Q and Zhu R-F 2005 *Int. Biodeterior. Biodegrad.* **55** 233–8
[12] Kardos L, Palkó G, Oláh J, Barkács K and Záray G 2009 *Water Sci. Technol.* **60** 957–64

[13] Mobarak-Qamsari E, Kasra-Kermanshahi R, Nosrati M and Amani T 2012 *Int. J. Environ. Res.* **6** 475–80
[14] Zhang P, Chen Y and Zhou Q 2009 *Water Res.* **43** 3735–42
[15] Jiang J, Zhang Y, Li K, Wang Q, Gong C and Li M 2013 *Bioresour. Technol.* **143** 525–30
[16] Lins P, Malin C, Wagner A O and Illmer P 2010 *FEMS Microbiol. Ecol.* **71** 469–78
[17] Lossie U and Pütz P 2008 *Practice Report Hach-Lange* http://tr.hach.com/asset-get.download.jsa?id=25593611361
[18] Rinzema A, van Lier J and Lettinga G 1988 *Enzyme Microb. Technol.* **10** 24–32
[19] Klinke H B, Ahring B K, Schmidt A S and Thomsen A B 2002 *Bioresour. Technol.* **82** 15–26
[20] Borja R, Martin A, Maestro R, Alba J and Fiestas J A 1992 *Process Biochem.* **27** 231–7
[21] Zhang C, Sanders J P M and Bruins M E 2014 *Biomass Bioenergy* **67** 466–72
[22] Yen H-W and Brune D E 2007 *Bioresour. Technol.* **98** 130–4
[23] Chen Y, Cheng J J and Creamer K S 2008 *Bioresour. Technol.* **99** 4044–64
[24] Otero M, Lobato A, Cuetos M J, Sánchez M E and Gómez X 2011 *Bioresour. Technol.* **102** 3404–10

Effect of particle size distribution and acid treated coal bottom ash on TSS and COD removal from textile effluent using fixed bed column

N L E Wahyuni, B Soeswanto*, H Akmal and N Puspita

Chemical Engineering Department, Politeknik Negeri Bandung, PO Box 1234, Bandung, Indonesia

*Email: bambang.soeswanto@yahoo.com

Abstract. Most wastewater treatment plant (WWTP) of textile industries in Bandung area have not operated optimally due to several reasons, such as high processing load and high operational cost. On the other hand, bottom ash which is a major pollutant generated from coal-fired boiler has good potential as adsorbent. In this paper bottom ash obtained from a textile industry in Bandung has been used to remove the organic contaminant from textile effluent in order to reduce the burden to WWTP. The ash was treated using acid solution prior to adsorption and filtration of waste, in order to avoid leaching of its metal content by wastewater and to enhance its adsorption capacity. The best contact time on COD removal and the adsorption capacity were obtained in batch processes. The effect of bed particle size to pressure drop, the influence of contact time, bed particle size on TSS filtration and COD removal were also studied using a fixed bed column. Acid treatment of ash was carried out using 0.25 M HCl solution. One gram of bottom ash was leached by 20 mL of acid solution to give 1.402 ppm of chromium effluent (below the threshold limit of 5 ppm). The highest adsorption of acid-treated bottom ash on Rhodamine B 233 was achieved on 2 h duration of adsorption, with adsorption capacity of 15.87 mg Rhodamine/g ash). Filtration of wastewater on fixed bed of finer grain (1.0625 mm) as well as coarser grain (1.200 mm) provided 55 % TSS reduction, 30% volumetric rate reduction, and 95% COD reduction in less than one hour.

1. Introduction

Textile industry is one of the most important sectors in Bandung and its surrounding areas. There are around 350 textile industries by 2016. The textile industries generally have difficulties in meeting wastewater discharge limits, particularly with regard to dissolved solids, pH, BOD, COD, heavy metals, and color of effluent. On the other hand, a large amount of bottom and fly ashes that are produced during combustion of coal in power plant requires large dumping sites for the safe disposal, due to the relatively low level of ash utilization rate.

Adsorption, a displacement phenomenon on a surface, is strongly influenced by particle size distribution and adsorbent porosity. Most of textile industries in West Java use chain grate boiler which operates at a relatively low combustion temperature that produces a fairly large amount of bottom ash. The bottom ash has a relatively high carbon content, estimated around 30%. Due to the heating process, there is an increase in carbon porosity. Therefore, it is potential to be used as an effective adsorbent. To reduce the load of textile wastewater treatment plant (WWTP), an initial treatment of waste can be conducted using coal bottom ash. The leaching studies of coal bottom ash indicate that the heavy metals in the bottom ash and fly ash are bound to different fractions with different strengths. From the environmental and utilization perspectives, heavy metals in ashes posed different levels of environmental contamination risk.

Therefore, the bottom ash needs to be extracted prior to utilization as an adsorbent in order to give the environmentally responsible release of wastewater. The heavy metals content of the effluent resulted from acid treatment should meet the threshold limit. Besides that, ash treatment with acid is necessary to improve the adsorption performance.

The effective size and uniformity of grain to characteristics of granular media treatment systems [1]. Many researchers had investigated the utilization of coal ash as a low-cost adsorbent and filter medium for uptaking of organic compound from waste effluents, including adsorption of reactive dyes [2], utilization of NaOH-activated coal bottom ash as Cd^{2+} ion adsorbent [3], utilization of bottom ash to adsorb heavy metals [4], organic pollutants[5]and dyes [6,7], utilization of coal ash as an adsorbent of organic matter in brackish water [8], and the utilization of bottom ash as a filter medium for textile wastewater treatment. Treatment of artificial textile dyes waste (methylene blue) were adsorbed using fly ash [9]. The use of coal ash as an adsorbent was chosen due to its high silica and alumina content. Coal ash is a low-cost adsorbent compared to other adsorbents, but the ash contains some heavy metal elements as trace elements such as Fe, Mn, Zn, Cu, Cr, Ni, Mg, Pb, Li, Co, Hg, Cd and As which are potentially extracted when contacted with acidic solutions. Therefore, coal ash is classified as a hazardous waste [10]. Different elements have different leaching behaviors due to differences in elemental properties and pH of the solution as well as leaching time [11]. The fly ash treated with acid solution increased the adsorption surface area by 60% [12].

The objectives of this paper were to study the effect of coal bottom ash particle size distribution on the extent of total suspended solid (TSS) removal in a fixed bed column and to investigate the effect of ash initial treatment on the COD removal of artificial wastewater using batch process and a fixed bed column.

2. Materials and methods

The bottom ash utilized as adsorbent and filter medium in this study originated from a textile industry in Cimahi, Province of West Java, Indonesia. The TSS content of artificial wastewater was prepared by mixing fine particles of bentonite with tap water. The COD content was prepared by dissolving Rhodamine B 233 in tap water.

2.1. Filtration for TSS removal

The filtration was conducted by means of an experimental seepage column (100 cm height and 5 cm inside diameter, made of Plexiglass), completed with pressure drop indicators, which were installed at different heights in the filter. The initial ash treatments were grinding and sizing to reduce the size from about 5 cm to 14- 20 mesh (1.40 – 0.85 mm). The ground ash then washed using aquadest at 80°C for 4 h with stirring, followed by filtration and drying at 100°C. Ash washing was conducted to reduce the water-soluble metal oxide content as well as to clean ash pores from the impurities. The artificial wastewater with TSS content of 460 ppm was fed from a water tank to the upper constant head controller, flow throughout the column and then discharged via the lower constant head controller (valve). The flow rate and hydraulic head along the filtration path were observed manually every 30-60 min. The effect of particle size on pressure drop, effluent flow rate reduction and TSS removal was investigated for eight hours of operation.

2.2. Adsorption studies

2.2.1. Adsorbent preparation. Trace element analysis on coal bottom ash showed that the chromium content a hazardous metal was quite high (110 ppm). Therefore, it was necessary to leach the chromium content to prevent the environmental effects. The leaching was performed by mixing the ash in 0.25 M HCl solution (with acid to ash ratio = 20: 1), with stirring for 4 h, followed by filtration and drying at 105°C for 12 h in an oven [13]. Acid treatment on ash was also intended to open the pores of ash which will increase adsorption capacity. The best contact time of COD removal was obtained for initial concentration of 750 ppm.

2.2.2. Isotherm adsorption. In order to identify the appropriate adsorption isotherm of COD, a series of batch adsorption experiments of artificial wastewater were performed using the best contact time obtained from the former experiment. Langmuir equilibrium isotherms were analyzed according to the experiment data and the adsorption capacity of the ash was estimated.

2.2.3. Column test. Fixed bed experiments were performed with constant bed height and best particle sizes obtained from the filtration experiments. As is known, adsorption is a transient (time dependence) process, the amount of material adsorbed within a bed depends both on position and time. As fluid enters the bed, it comes in contact with the upper layer of absorbent, then solute filling up the available sites. The adsorbent near the entrance is saturated and the fluid penetrates farther into the bed before all solute is removed. Thus, the active region shifts down through the bed as time goes on. The breakpoint was obtained at the effluent's TSS of 200 ppm. Analysis of TSS was conducted by means of SNI 06-6989.3-2004. Analysis of COD was conducted by means of SNI 6989.73-2009 method. Chromium concentration in the effluent was measured with AAS GBC Avanta Ver. 2.02.

3. Results and discussion

3.1. Filtration

The experiment to study deep bed filtration was based on a visual observation of pressure drop, measuring the flow rate and TSS of the effluent.

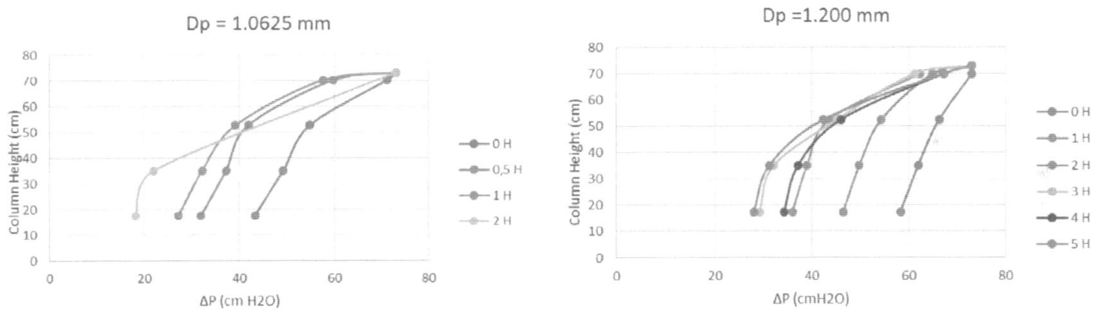

Figure 1. Progress of the filter bed resistance in time.

From Figure 1, it is found that over the operation time the pressure drops of both filter beds getting bigger with time. The bed filter with grain diameter of 1.200 mm gave lower pressure drop and longer period of time than that of 1.0625 mm. The bed with finer grain showed a very rapid increase in head loss. The pressure dropped along the column height fit smoothly with curve, indicated that clogging occurred evenly in the whole bed.

The filtration was intended as a preliminary treatment to WWTP, which effluent will be further treated. Therefore, the effluent of filtration was considered to be satisfactory with TSS content of 200 ppm. The effect of contact time on velocity of effluent and TSS level were presented on Figure 2a and 2b, respectively.

Based on Figure 2a and 2b, the bed with grain size of 1.200 mm was considered more satisfactory than the bed with grain size of 1.0625 mm, which reduced TSS value from 430 to 200 ppm and 100 ppm during one hour and two hours of contact time (one hour service run), with 30% effluent velocity reduction (from 36.67 to 24.42 m^3/m^2.day).This filtration belongs to the medium speed (quick sand filter: 100-475 m^3/m^2.day; slow sand filter: 4 - 8 m^3/m^2.day).

IOP Conf. Series: Earth and Environmental Science **160** (2018) 012015 doi:10.1088/1755-1315/160/1/012015

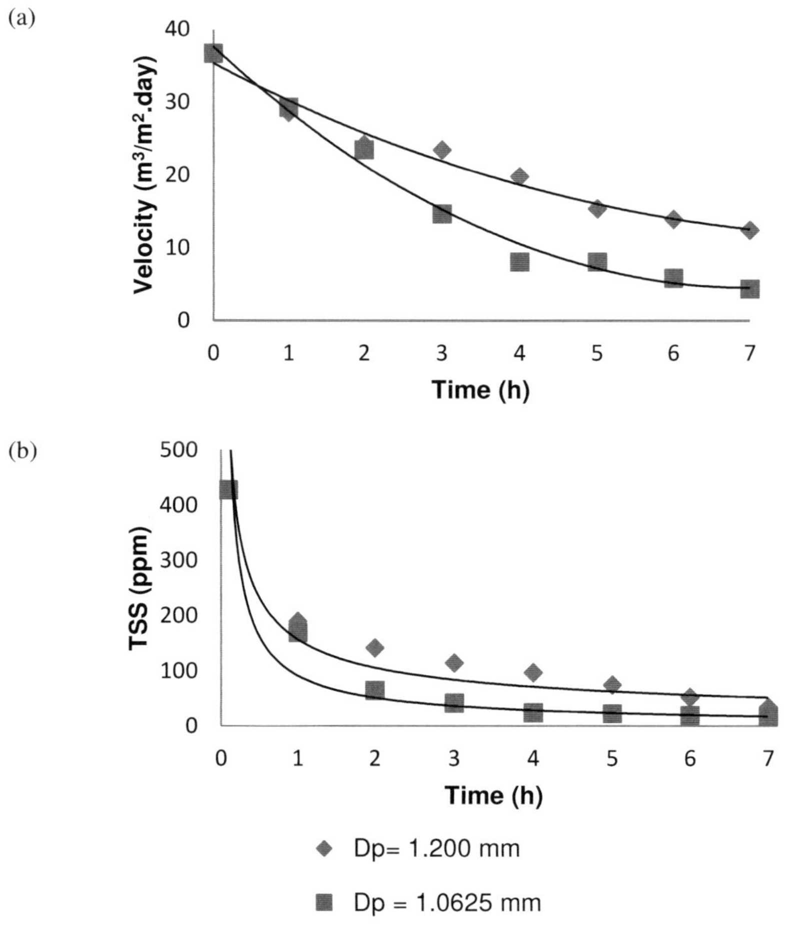

Figure 2. Effect of time on (a) velocity and (b) TSS of effluent.

3.2. Adsorption

3.2.1. Adsorbent treatment and contact time. Acid treatment of 1.200 mm bottom ash using 0.25M HCl solution with ratio of 1 g ash to 20 mL of acid leached the highest heavy metal with chromium by 74.52%, which gave effluent concentration of 1.4016 ppm (below the threshold limit of 5 ppm [14]).

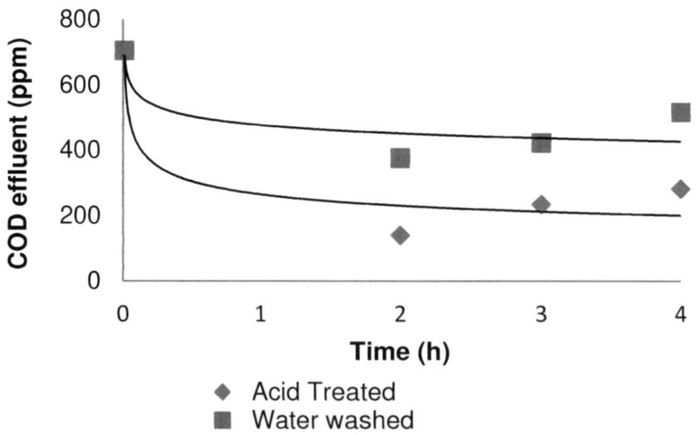

Figure 3. COD removal in batch adsorption.

The determination of best contact time was performed by means of batch adsorption for COD removal. The result is presented in Figure 3. Based on Figure 3, the best contact time was two hours. The figure also showed that ash treatment using solution of 0.25 M HCl enhanced the COD removal from wastewater, almost 100% higher than that of water-washed ash of which 68 and 34%, respectively.

The enhancement of treated ash adsorption was in line with the work of Bada and Potgieter-Vermaak [12] who found that the fly ash treated with acid solution has increased the adsorption surface area by 60% (due to increasing pore volume). The appearance of the liquid underwent adsorption for longer time than two hours was cloudy, indicated that for contact time longer than two hours, the adsorbent was eroded resulting in the adsorbate release from the adsorbent.

3.2.2. Isotherm adsorption. The adsorption equilibrium data plot in the form of an isotherm adsorption is presented in Figure 4.

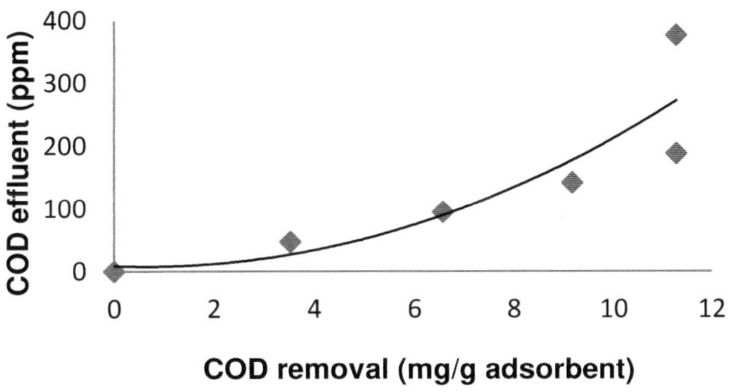

Figure 4. Isotherm adsorption of COD removal.

The shape of isotherm adsorption curve was categorized as favorable, which permitted higher solid loadings at lower solution concentrations. The mathematical interpretation of the adsorption isotherms was studied using the Langmuir model. For the Langmuir model linearization gives:

$$\frac{C_e}{q_e} = \frac{1}{K.Q_0} + \frac{C_e}{Q_0}$$

q_e = Concentration of COD in mg/g of ash; C_o = Initial concentration of COD in solution in ppm; C_e= Final concentration of COD in solution in mg/L.

The plot of C_e/q_e versus C_e is presented on Figure 5, which gives a straight line with intercept of $1/(K.Q_0)$ and slope of $1/Q_0$.

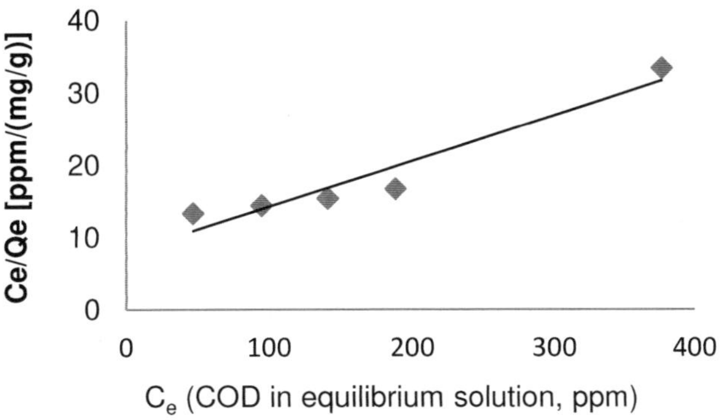

Figure 5. Langmuir isotherm adsorption.

Figure 5 shows that the isotherm adsorption was linear with R^2 value of 0.924, indicating the applicability of Langmuir model. The adsorption capacity was found from the slope of the curve to be 15.873 mg/g ash.

3.2.3. Breakthrough curve and evaluation of continuous adsorption. Evaluation of adsorption performance in the continuous process was conducted by observing the emerging of unabsorbed solute in the effluent (breakthrough curve). The plot is presented on Figure 6. Artificial wastewater with COD of 750 ppm and TSS value of 430 ppm was prepared as feed mixture in order to examine the effect of TSS content on the adsorption.

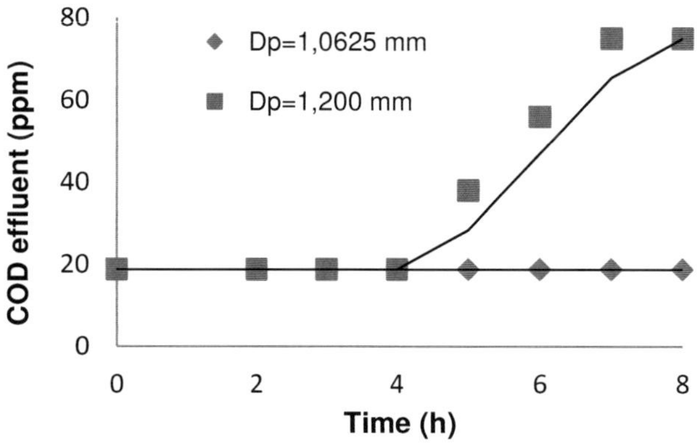

Figure 6. Breakthrough curve of adsorption.

Figure 6 shows that in the bed with grain diameter of 1.200 mm the unabsorbed solute emerged in the effluent at four hours service time, and increased with time. As the time went on the COD of effluent increased, but the process did not continue for more than 8 h because the effluent flow rate kept decreasing. Therefore, the breakthrough curve was not completed yet. However the bed with grain diameter of 1.06 mm, the effluent remained clean until 8 h of operation.

4. Conclusion

The results showed that the bed column with coarse grain provided better TSS reduction than finer grain. TSS removal was achieved at relatively short service run with the medium speed filtration. The COD removal with acid-treated coal bottom ash inceased by100%. In column with coarse grain, the unabsorbed solute started to emerge in the effluent at 4 h of service time and increased with time. Whilst the bed with fine grain resulted clean effluent until 8 h of operation.

References

[1] Eliasson J 2002 *Sand/Media Specifications, Rule Development Committee Issue Research Report, Wastewater Management Program* Washington State Department of Health
[2] Jarusiripot C 2014 *Procedia Chem.* **9** 121-30
[3] Wardani R K 2011 *Pemanfaatan Abu Bawah Batu Bara (Bottom Ash) Teraktivasi Sebagai Adsorben Ion Logam* Cd^{2+} Bachelor Thesis (Surabaya, Indonesia: Airlangga University)
[4] Ayala J, Blanco F, García P, Rodriguez P and Sancho J 1988 *Fuel* **77** 1147–54
[5] Aksu Z and Yener J 2001*Waste Manag.* **21** 695–702
[6] Sari PI 2010 *Adsorpsi Methilen Biru Dengan Abu Dasar PTIPMOMI Paiton Jawa Timur dan Zeolit Berkarbon* Bachelor Thesis (Surabaya, Indonesia: Institut Teknologi Sepuluh November)
[7] Wang S, Boyjoo Y, Choueib A and Zhu Z H 2005 *Water Res.* **39** 129-38
[8] Sugiharta I G B 2011 *Utilization of Coal Bottom Ash as Adsorbent of Organic Material in Brackish Water* Bachelor Thesis (Surabaya, Indonesia: UPN Veteran East Java)
[9] Das R, Saha P and Sinha I 2015 *Int. J. Emerging Technol. Adv. Eng.* **5** 543-8

[10] Juda-Rezler K and Kowalczyk D 2013 *Pol. J. Environ. Stud.* **22** 25-40
[11] Kishor P, Ghosh A K and Kumar D 2010 *Asian J. Agric. Res.* **4** 1-14
[12] Bada S O and Potgieter-Vermaak S 2008 *Leonardo Electron. J. Practices Technol.* **12** **7**37-48
[13] Lokeshappa B and Dikshit A K 2012 *Resour. Environ.* **2** 1-8
[14] The Republic of Indonesia Minister of Environment Decree No. 5 Year of 2014 Wastewater Quality Standard

IOP Conf. Series: Earth and Environmental Science **160** (2018) 012016 doi:10.1088/1755-1315/160/1/012016

Nickel removal from electroplating wastewater using electrocoagulation

Djaenudin, Muchlis and Ardeniswan

Research Unit for Clean Technology, Indonesian Institute of Sciences (LIPI)
Komplek LIPI Bandung, Jl. Sangkuriang, Gedung 50, Bandung 40135, Indonesia

E-mail: djae004@lipi.go.id; aburizki@yahoo.com

Abstract. Electrocoagulation processes for reducing nickel concentration in electroplating wastewater had been carried out in laboratory scale experiments. The experiments were conducted in a 500 mL batch reactor with dual electrodes: aluminum as the anode and copper as the cathode with electroplating wastewater at pH 7 and initial nickel concentration of 10,55 ppm. Parameters of electrocoagulation process such as removal efficiency, specific energy and produced precipitate were determined and assessed under various DC voltages (5, 7.5 and 10 V) and various operation time (30, 60 and 90 min) as variable of experiment. The results show that the optimum condition of electrocoagulation process was obtained at 5 VDC and for 76.5 min.

1. Introduction

The development of electroplating industry in Indonesia grows rapidly. Large amount of wastewater is produced daily by the electroplating industry and provide negative impact in environmental quality. Wastewater from electroplating industry usually contain heavy metals that may endanger living beings' health and environment [1–4]. Moreover, the water pollution by heavy metals cause serious problems to human being due to bioaccumulation effects in aqueous environment [5].

Most of harmful elements in electroplating wastewater contain heavy metals such as nickel [1]. Nickel is the main pollutant in electroplating wastewater due to only 30-40% of nickel that can be utilized into product, while the rest is released into water body as wastewater effluent [1]. The Decree of the Minister of Environment Indonesia No. 5 in 2014 regulates the maximum nickel concentration in effluent released into water body must be less than 1 ppm [6]. For fulfilling the standard quality, the electroplating wastewater effluent must be treated by appropriate wastewater treatment technology. Various techniques for heavy metal removal from wastewater are available such as ion exchange, adsorption, filtration, coagulation and electrocoagulation as alternative solutions [1–5,7–15].

Electrocoagulation is an emerging technology in water and wastewater treatment that has been successful in dealing with organic and inorganic contaminants. It has gained immense attention due to its ease of distribution and moderate required amount of chemicals [4,7–11,13]. The technique involves dissolution of the sacrificial anodes that produce active metal hydroxide as strong coagulant. Then the removal of particles from wastewater proceed through destabilizing/neutralizing the repulsive forces that keep the contaminants suspended in water. When the repulsive forces are neutralized, the suspended contaminants will form large particles that can settle down for easier separation from wastewater. The success of this technology will depend on its ability to satisfy commercial criteria such as minimizing operational and maintenance costs as well as achieving profit on the investment made in the shortest time possible. The major challenge that the technology is likely to encounter is that of high electricity consumption while treating the wastewater as it directly affects the operating costs [4,7–11,13]. Electrocoagulation process is influenced by operating parameters such as applied current density, electrodes material and configuration, type of electrical connection, pH and conductivity of the solution, and mixing state [4,7–11,13].

In this study, the influences of electrocoagulation process variables such as DC voltages and operation time were investigated in laboratory scale batch reactor. Effluent of an electroplating plant with typical high nickel concentration was used in the experiments. The optimum condition was expected to give a higher removal efficiency, lower specific energy and lower produced precipitate.

2. Materials and Method

2.1. Electroplating wastewater

The wastewater was collected from the effluent of an electroplating plant in Bandung City, West Java, Indonesia. The wastewater was characterized for pH, conductivity and nickel concentration (Table 1).

Table 1. Characteristics of electroplating wastewater for the experiment.

Characteristics	Value	Quality standard [6]
Nickel concentration (ppm)	10,523	1
TSS (ppm)	0.36	20
pH	3.4	6-9
Conductivity (mS/cm)	16	-
Color	Light green	-

Prior to experiment, the wastewater was filtered using a screen filter to remove suspended solids and the pH was adjusted to neutral condition by NaOH addition.

2.2. Electrocoagulation procedure

The setup of electrocoagulation equipment system is shown in Figure 1 and Figure 2. The electrocoagulation reactor was constructed from Pyrex a borosilicate glass beaker with 600 mL capacity, having dimensions of 90 mm outer diameter and 124 mm height. The total volume of wastewater in each experiment was approximately 500 mL. Aluminum (Al) plate was used as anode and copper (Cu) plates was used as cathode. Both plates have 45 mm height × 75 mm width × 3 mm thickness in dimension and were arranged in monopolar configuration with the distance between plates was fixed at approximately 50 mm. The current was maintained constant by means of a precision DC power supply characterized by the ranges 0–6 A for current and 0–10 V for voltage.

The experimental range and levels of independent process variable were DC voltage (5, 7.5 and 10 V) and operation time (30, 60 and 90 min), as shown in Table 2. The wastewater was agitated with a magnetic stirrer at 200 rpm during electrocoagulation. Electrodes were sanded and washed with dilute HCl between experiments.

To follow the progress of the treatment, samples of 10 mL were taken from the electrocoagulation cell at the end of the experiment. The samples were then filtered (0.45 μm) to eliminate sludge formed during electrolysis. The filtrate from each experiment were analyzed using ICP-OES instrument (Agilent 240FS AA) for determining the residual nickel concentration. The produced precipitate and electrode consumption were analyzed using gravimetric method [16]. The pH reduction and the specific energy consumption were also analyzed during each process.

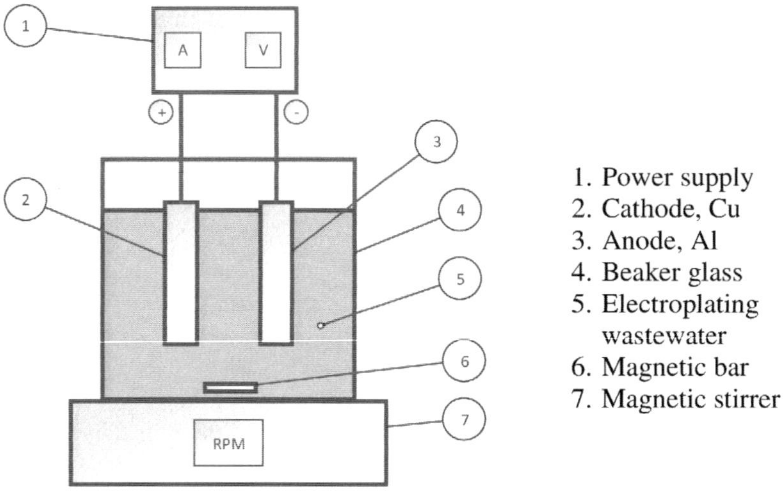

Figure 1. Schematic of electrocoagulation equipment system.

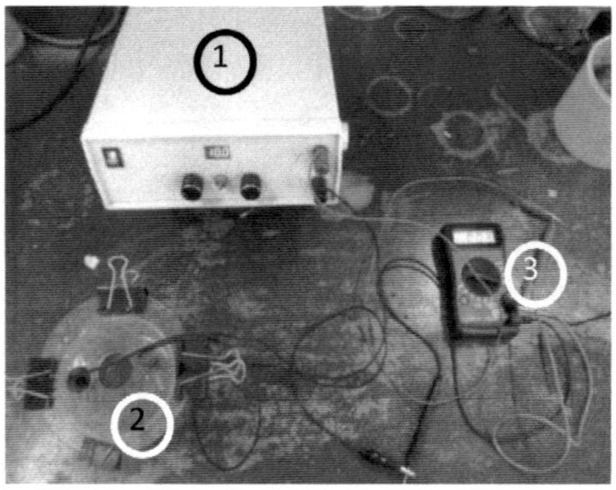

Figure 2. Electrocoagulation apparatus. A rectifier (1) acted as an AC-to-DC converter and an electric voltage controller; a coagulation reactor (2) served as a wastewater treatment plant; an AVOmeter (3) acted as an electric current meter and connecting wires of such devices.

Table 2. Experimental range and levels of independent process variable.

Process variable	Range and Value		
	Low	Middle	High
DC voltage (V)	5	7.5	10
Operation time (min)	30	60	90

2.3. *Calculations*

Nickel removal efficiency (%) was calculated based on initial nickel concentration (C_0, ppm) and residual nickel concentration (C_r):

$$Removal = \frac{C_0 - C_r}{C_0} x100\% \tag{1}$$

The high electricity consumption while treating the wastewater with electrocoagulation process is a major challenge and directly affects the operating costs, thus the electrical energy consumption is a very important economical parameter. The electrical energy consumption was calculated using the following equation:

$$E = \frac{U.I.t}{V} \tag{2}$$

where E was the energy consumption (kWh/m^3), U was the applied voltage (V), I was the current intensity (A), t was the electrocoagulation time (h) and V was the volume of the treated wastewater (L).

The amount of electrode dissolved was calculated theoretically by using Faraday's law:

$$C = \frac{I.t.M}{Z.F.V} \quad (3)$$

where C was the aluminum concentration in the electrolytic cell (g/L), I was the current intensity (A), t was the electrocoagulation time (s), M was the molecular weight of the anode (g/mol), Z was the chemical equivalence, F was the Faraday constant (96500 C/mol) and V was the volume of the treated wastewater (L) [9].

3. Results and Discussion

Electrocoagulation process was carried out to determine the effect of variations in voltage and time on the concentration of nickel (Figure 3).

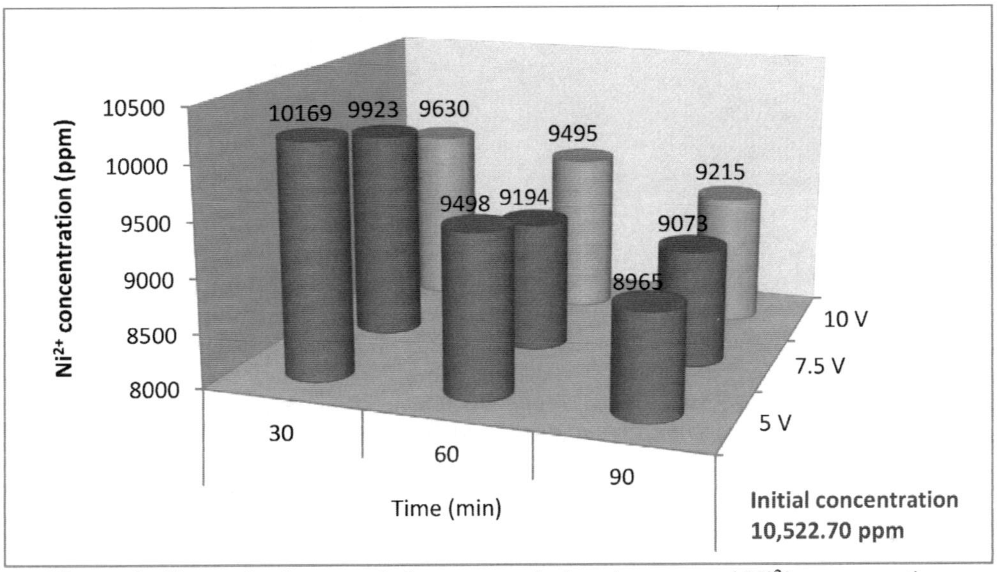

Figure 3. Chart of voltage and electrocoagulation time toward Ni^{2+} concentration.

Analysis results of Ni^{2+} concentration in the samples shows a reduction that was not significant (Figure 3). Overall, the concentration of Ni^{2+} of 10,522.70 ppm in the initial sample decreased until it reached the smallest concentration was of 8,965.15 ppm at 5 V for 90 min.

Pahlevi *et al* [4] suggested that decrease levels could be caused by the formation of precipitate and floating flocculants which indicated the movement of Ni^{2+} ions with an orbit helix-shaped around the electrode plate, so there was a tendency of Ni^{2+} ions to adhere into the whole electrode plate (cathode). The graph above also indicates the reduction of nickel levels that were influenced by the supply of Al^{3+} cations into the wastewater. This was caused by the more Al^{3+} cations supplied so more cations that could absorb or bind pollutants (in this case was a heavy metal compounds named nickel). With a large of anode's weight (Al) scraped would influence into reduction of nickel content in wastewater [8,9].

Figure 4 shows the nickel removal from the wastewater. Nickel removal by electrocoagulation in this study was not obtained perfectly, where the maximum reduction of nickel content only reached 14.8% at 5 V for 90 min. Relationships with reduction of nickel concentration are shown in Figure 4. The amount of precipitate formed, changes in pH, and the amount of the anode scraped would also indicate a decrease of the dissolved nickel in electroplating waste.

The oxidation reactions at the anode occurred as follows [18]:

$$2H_2O_{(l)} \rightarrow 4H^+_{(aq)} + O_{2(g)} + 4e^-$$
$$Al_{(s)} \rightarrow Al^{3+}_{(aq)} + 3e^-$$

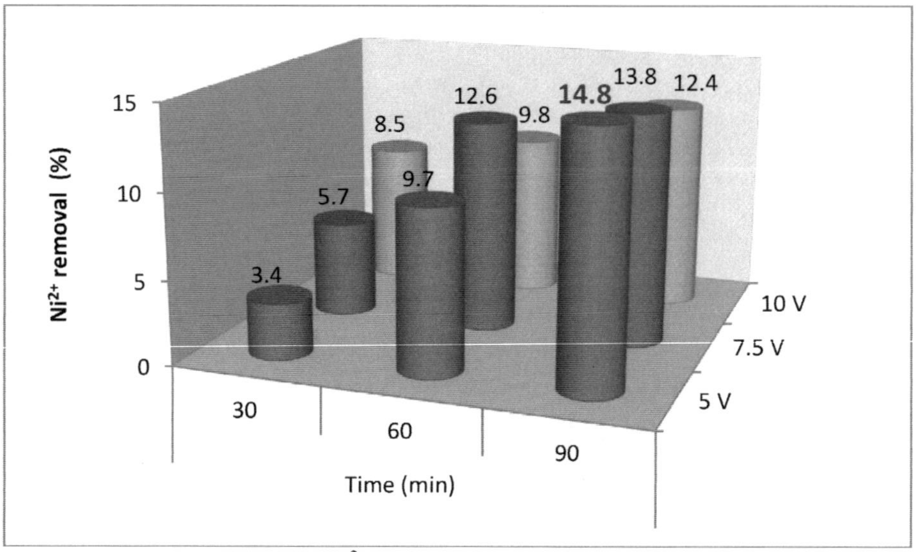

Figure 4. Chart of Ni^{2+} removal in electrocoagulation process.

At the cathode, the Al^{3+} ions were not reduced because of E°red Al^{3+} < E°red H_2O, so there would be a reduction reaction of water that produced H_2 gas and hydroxide ions.

$$2H_2O_{(l)} + 2e^- \rightarrow 2OH^-_{(aq)} + H_{2(g)}$$

The metal ions would be reduced to their metal and present on the electrode surface.

$$Cu^{2+}_{(aq)} + 2e- \rightarrow Cu_{(s)}$$

Based on the above reaction Al^{3+} ions would be released from the plate electrode (anode). The Al^{3+} ion would react with the hydroxide ions from the cathode to form $Al(OH)_3$ precipitate according to the following reaction:

$$Al^{3+}_{(aq)} + 3OH^-_{(aq)} \rightarrow Al(OH)_{3(s)}$$

As shown in above reactions, during electrocoagulation, $Al(OH)_3$ was produced by the aluminum anode and the hydroxide ions (OH^-) derived from the H_2O molecules. The free OH^- also would form metal hydroxide compounds ((M(OH)x) that were insoluble in water [8,9]. In addition, $Al(OH)_3$ as a coagulant would further adsorb the existing pollutants into their molecular cavities to form a complex metal oxide compound (MAl_2O_4) and insoluble in water [17].

As also shown in above reactions, O_2 gas and H_2 gas would be produced in the anode and cathode, respectively. $Al(OH)_3$ then formed flocs that interacted with the gas bubbles to generate low-density flocs that would move to the surface [8,9]. The flocks formed will have a relatively small size so that the flock will increase its size along with the electrocoagulation process time [9]. This matter plays a role in lifting the contaminant of the solution onto the surface of the reactor, which is usually called the flotation process.

The amount of precipitate formed in the electrocoagulation process is shown in Figure 5. The precipitate was a mixture of various compounds, consisted of Ni^{2+} that was coagulated by $Al(OH)_3$. The maximum precipitate weight of this process, 0.185 g, was generated at 5 V and 90 min. This maximum precipitate weight corresponded to the decrease in dissolved nickel content, which also reached the optimum point at 5 V and 90 min.

Figure 6 shows the pH value of the solution at the end of the experiment. There was a decrease in the pH value of the solution from the initial pH 7. Usually the variation of the voltage does not significantly affect the final pH of the experiment, whereas the experimental time greatly affects the final pH of the experiment. In our experiment, a significant decrease of pH occurred in almost every experimental variation, where the initial pH of waste (7) decreased to pH 4-5. The reaction that occurred in the anode caused the increase of H^+ ions. Furthermore, metal ions in the solution would bind the OH^- ions. Therefore, there was an excess of H^+ ions and the pH of decreased.

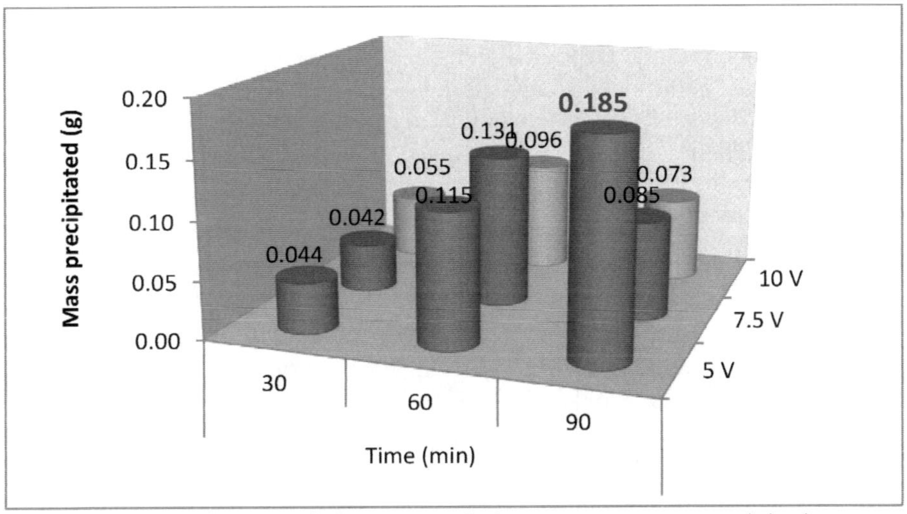

Figure 5. Influence of electrocoagulation parameters on precipitation.

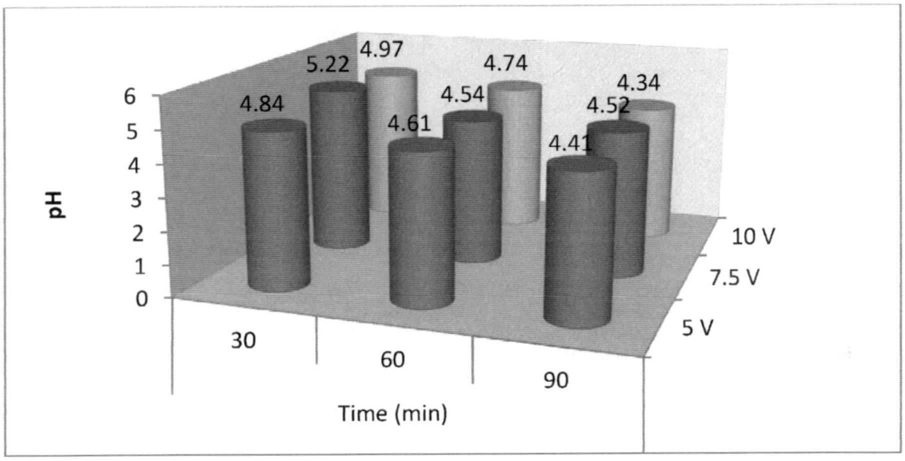

Figure 6. Final pH on electrocoagulation process.

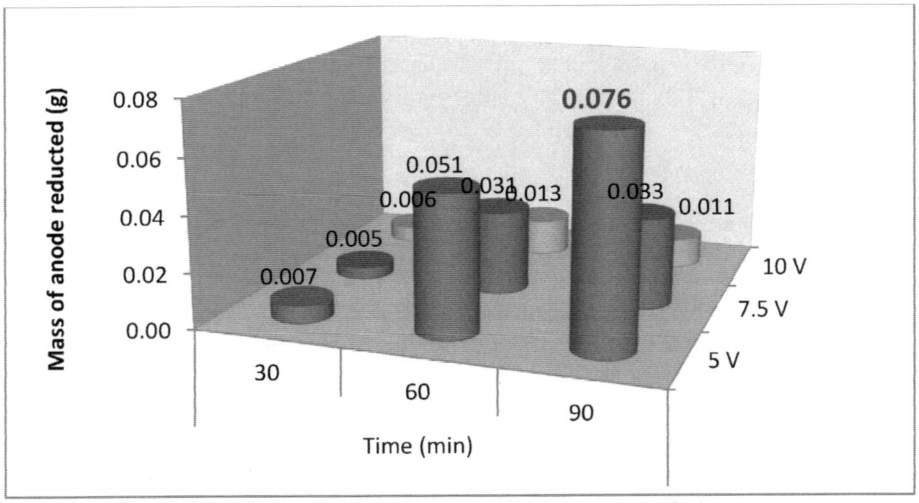

Figure 7. Weight of scraped anode on electrocoagulation process.

The magnitude of the eroded anode was affected by the voltage and electrocoagulation time as shown in Figure 7. The total mass of reduced aluminum anode reached a maximum of 0.076 g at 5 V and 90 min. The pH changes in the electrocoagulation process also influenced the amount of eroded Al and the formation of $Al(OH)_3$ precipitate which indicated a decrease in dissolved nickel content therein.

Specific energy is the energy required per kg of nickel removed from the solution. The smaller the specific energy, the smaller the energy used to decrease the dissolved nickel content. This would affect the cost efficiency of the process.

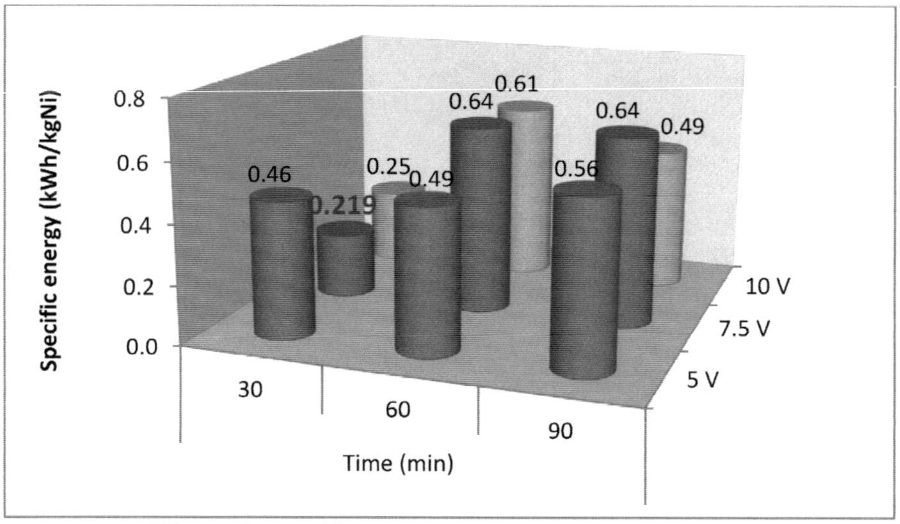

Figure 8. Specific energy for electrocoagulation process.

Figure 8 shows the specific energy required for each voltage and time variation. The smallest specific energy was achieved at 7.5 V and 30 min, which means that the process at these conditions would bear minimum cost with specific results. There was a difference between the optimum point of specific energy with decreased levels of nickel and the precipitate produced in the electrocoagulation process.

4. Conclusion

The research concluded that the percentage of reduction of dissolved nickel concentration in the sample by electrocoagulation reached 14.8% from 10,553.70 ppm to 8,965.15 ppm. It is indicated that the electrocoagulation process can be used to reduce dissolved nickel. However, the experimental results in this study have not succeeded in reducing the concentration of soluble nickel until reaches the quality standard as stated in the regulation of Environment Minister No. 5 year 2014.

Acknowledgments

The authors would like to acknowledge DIPA program of Research Unit for Clean Technology for funding and equipment support. We also thank to Rihlatul Adni for sampling and analysis support.

References

[1] Dermentzis K 2010 *J. Hazard. Mater.* **173** 647–52
[2] Herlina F 2016 *Info Tek.* **14** 171–85
[3] Nurhasni N, Salimin Z and Nurfitriyani I 2013 *J. Kim. Val.* **3** 41–7
[4] Pahlevi M N, Mirwan A and Triwulandari R 2012 *Konversi* **1** 45–50
[5] Avessa I, Yusuf B and Alimuddin A 2016 *J. Kim. Mulawarman* **14** 7–12
[6] Minister of Environment, Republic of Indonesia Peraturan Menteri Lingkungan Hidup Republik Indonesia No. 5 Tahun 2014 tentang Baku Mutu Air Limbah

[7] Wahyulis N C, Ulfin I and Harmami H 2014 *J. Sains Seni ITS* **3** C9–11
[8] Martínez-Huitle C A and Brillas E 2009 *Appl. Catal. B Environ.* **87** 105–45
[9] Moussa D T, El-Naas M H, Nasser M and Al-Marri M J 2017 *J. Environ. Manage.* **186** 24–41
[10] Purwaningsih I 2008 *Pengolahan limbah cair industri batik CV. Batik Indah Raradjonggrang Yogyakarta dengan Metode Elektrokoagulasi Ditinjau dari Parameter Chemical Oxygen Demand (COD) dan Warna* Bachelor Thesis (Yogyakarta, Indonesia: Universitas Islam Indonesia)
[11] Sahlan L, Radinta S, Kholisoh S D and Mahargiani T 2016 *Seminar Nas. Tek. Kim. Kejuangan* I13
[12] Suprihatin and Indrasti N S 2010 *Makara J. Sci.* **14** 44–50
[13] Un U T and Öcal E 2015 *Int. J. Environ. Sci. Dev.* **6** 425–9
[14] Tumin N D, Chuah A L, Zawani Z and Rashid S A 2008 *J. Eng. Sci. Technol.* **3** 180-9
[15] Wilyanda W, Yelmida Y and Chairul C 2015 *J. Online Mhs. JOM Bid. Tek. Sains* **2** 1–10
[16] Svehla G, Setiono L, Pudjaatmaka and Hadyana A 1985 *Vogel: Buku Teks Analisis Anorganik Kualitatif Makro dan Semimikro* (Jakarta: Kalman Media Pusaka)
[17] Huheey J E, Keiter E A, Keiter R L and Medhi O K 2006 *Inorganic Chemistry: Principles of Structure and Reactivity* (Delhi: Pearson Education India)
[18] Budi E H, Patterson L B and Parichy D M 2011 *PLOS Genet.* **7** e1002044

IOP Conf. Series: Earth and Environmental Science **160** (2018) 012017 doi:10.1088/1755-1315/160/1/012017

Isotherm adsorption studies of Ni(II) ion removal from aqueous solutions by modified carboxymethyl cellulose hydrogel

L Anah[*] and N Astrini
Research Unit for Clean Technology, Indonesian Institute of Sciences (LIPI)
Komplek LIPI Bandung, Jalan Sangkuriang, Gedung 50, Bandung, 40135, Indonesia

[*]Email: likanah@yahoo.com

Abstract. This paper reports the removal of nickel ions from aqueous solutions using carboxymethyl cellulose-graft-poly(acrylic acid) CMC-g-PAA hydrogel as an adsorbent. The hydrogel was prepared through graft copolymerization of acrylic acid monomer onto a carboxymethyl cellulose backbone in the presence of *N,N'*-methylenebisacrylamide as a crosslinking-agent using benzoyl peroxide initiator system. Isotherm for the adsorption of nickel on CMC-g-PAA hydrogel was developed and the equilibrium data fitted to the Langmuir and Freundlich isotherm models. The result showed that at pH of 5 and shaker speed of 100 rpm, Freundlich model was better fitted than Langmuir model, of which R^2 = 0.999 and R^2 = 0.935, respectively. The constant of Langmuir model was K_L value = 0.003 and q_{max} = 1000 mg/g, whereas the constant of Freundlich model n = 1.2 and K_F = 4.614. The CMC-g-PAA hydrogel before and after Ni(II) adsorption were characterized using scanning electron microscopy (SEM) and Fourier transform infrared (FTIR).

1. Introduction

Environmental pollution by toxic metals arises from industries, agricultural effluents and waste disposal from various sources. Many industries such as metal plating facilities, mining operations and tanneries discharge waste containing heavy metal ions.

As pollutants, heavy metals were intensively studied due to their significance from the point of view of persistence and toxicity. These toxic metals can cause accumulative poisoning, cancer, and brain damage when present at above the tolerance levels. The agencies for the environmental monitoring have set permissible limits for heavy metals levels in drinking water because of their harmful effects. The removal and rapid decontamination of heavy metals such as Cd, Pb, Cu, Ni, Hg become very important for the environmental remediation [1]. The World Health Organization has recommended strict controls on the percentages of various heavy metals in effluent waters [2].

Hydrogels are water-swollen crosslinked network of hydrophilic polymer. Due to the hydrophilic groups in their backbone, hydrogels can imbibe large quantities of water and swells. The degree of hydrogel swelling is related to the degree of crosslinker, ionic strength of the solution, temperature, pH, and the presence and extent of hydrophilic groups such as –OH, -COOH, $-NH_2$, $-CONH_2$, and SO_3H. These functional groups in the hydrogel networks can be utilized for the toxic metal ion removal. Hydrogels can also be modified with new functionalities or prepared as composites with clays and with other materials to increase the metal ion absorption capacities [3–6].

Cellulose is renewable, and due to its abundance in nature it affords material for cost-effective hydrogel and ion exchange technologies. Cellulose and its derivatives have been used in metal ion adsorption. Sodium carboxymethyl cellulose (CMC) is an important biopolymer due to its high degree of solubility in water, biocompatibility, nontoxicity, and low cost [7]. Grafted cellulose products, when used as ion exchangers, have advantages over conventional ion exchangers due to chemical resistance, low cost of preparation, and they also offer considerable hydrophilic surface area. Acrylic acid and acrylamide monomers have been successively grafted onto a cellulose backbone. Grafting is a simple technique to incorporate desired active functional groups on the backbone of a polymer for absorption metal ions [8–10].

Adsorption techniques have been proved to be an excellent method for waste water, offering advantages over conventional processes. It is now recognized that adsorption using low cost-adsorbents is an effective and economic method for the removal of heavy metals in waste water. However, low-cost adsorbents with high adsorption capacities are still under development to reduce the adsorbent dose and minimize disposal problems. Much attention has recently been focused on biopolymers as bioadsorbent. They are naturally produced by all living organisms. Bioadsorbents are a useful option since they have no adverse impact on human or environmental health [11].

Isotherm adsorption models have been used in waste stream treatment to predict the ability of a certain adsorbent to remove a pollutant down to a specific discharge value. When a mass of adsorbent and waste stream are in contact for a sufficiently long time, equilibrium between the amount of pollutant adsorbed and the amount remaining in solution will develop. The Langmuir and Freundlich isotherm models are only applicable to batch adsorption systems where sufficient time is provided to allow equilibrium occurs between the pollutant in solution and the pollutant adsorbed on the media. During the flow through the adsorbent, a large amount of pollutants is expected to come into contact with active surface sites and thus be retained on the surface of the adsorbing media [12]. The Langmuir and Freundlich equations are most widely used. The Langmuir model assumes that the adsorption of metal ions is a process occurring on a homogenous surface without any interaction between adsorbed ions. The adsorption proceeds until a complete monolayer is formed. The Freundlich model is based on adsorption on a heterogeneous surface [1].

The aim of this work was to predict the ability of a certain adsorbent to remove a pollutant down to a specific discharge value by the Langmuir and Freundlich isotherm adsorption models.

2. Experimental

2.1. Material

Analytical reagent grade of nickel nitrate ($N_i(NO_3)_2.6H_2O$) was used as a source of N_i(II). Sodium hydroxide (NaOH) and hydrochloric acid (HCl) were used pH adjustment. All these chemical were purchased from E-Merck Company Inc., Germany and used without further purification. Deionized water was used in this work. Hydrogel CMC-g-PAA was used as the adsorbent. *N,N'*-methylenebisacrylamide was used as a crosslink agent whereas benzoyl peroxide was used as an initiator.

2.2. Method

2.2.1. Preparation of CMC-g-PAA hydrogel. The graft copolymers CMC-g-PAA were synthesized by free radical and solution polymerization. CMC was dissolved in distilled water then stirred and purged by passing nitrogen gas. The solution was slowly heated up to the desired temperature then added with benzoyl peroxide (BPO) initiator and methylenbisacrylamide (MBA) crosslinker agent. The reaction was continued until the desired time and process of polymerization was achieved. The hydrogel adsorbent was oven-dried and washed with distilled water and acetone. This adsorbent was available from a previous study [13].

2.2.2. Nickel ions sorption. The experiments were carried out in 250 mL volumetric flasks, each using 100 mg of CMC-g-PAA hydrogel and 100 mL of 10 mg/L nickel-rich waste water. The flasks were placed on a Gallenkamp Orbital Incubator Shaker running at 300 rpm, 30°C for 10 h. Adsorption of nickel ion was determined by using Atomic Absorption Spectrophotometer (AAS) Varian Type.

The amount of Ni^{2+} ions adsorbed by the CMC-g-PAA hydrogel adsorbent (Q_e) was calculated for each run by the following equation:

$$Q_e = \frac{(C_0 - C_e)V}{W} \tag{1}$$

where Q_e (mg/g) was the adsorption capacity; C_0 (mg/L) was the initial concentration of Ni^{2+}; C_e (mg/L) was the equilibrium concentration of Ni^{2+}; V (L) was the volume of the solution of Ni(II), and W (g) was the weight of adsorbent added [14, 15].

2.2.3. Equilibrium isotherm.

2.2.3.1. The Langmuir isotherm model. The Langmuir isotherm model equation was:

$$q = q_{max}\left(K_L \frac{C_e}{1 + K_L\, C_e}\right) \tag{2}$$

where q (mg/g) was the amount of heavy metal ions adsorbed onto the unit mass of the polymeric beads; K_L was the Langmuir equilibrium constant which is related to the affinity of binding sites; C_{eq} was the equilibrium Ni^{2+} ion concentration, and q_{max} was the maximum adsorption capacity (theoretical monolayer saturation capacity) [16].

The main characteristics of the Langmuir equation, constants K_L and q_{max}, could be determined from a linearized form of the Langmuir equation as follows:

$$\frac{C_e}{q_e} = \frac{1}{q_{max}\ K_L} + \frac{C_e}{q_{max}} \tag{3}$$

Therefore, a plot of C_e/q versus C_e gave a straight line with slope of $1/qmax$ and intercept of $1/(K_L.q_{max})$.

The Langmuir isotherm was used to describe observed sorption phenomena. The Langmuir isotherm applies to adsorption on completely homogenous surfaces with negligible interaction between adsorbed molecules [17].

2.2.3.2. The Freundlich isotherm model. The Freundlich isotherm model was based on adsorption on a heterogeneous surface, developed an empirical equation [1]:

$$q_e = KC_e^{1/n} \tag{4}$$

where q_e is the adsorption (mg/g); C_e was the concentration of adsorbate in the solution (mg/L); K_F and n were empirical constants that are characteristics of the system, indicating the adsorption capacity and the adsorption intensity, respectively. The above equation can be linearized as the following form, also used to confirm the applicability of the model [1]:

$$ln\, q_e = ln\, K + \frac{1}{n} ln\, C_e \tag{5}$$

where K_F was the measure of sorption capacity, and $1/n$ was the sorption intensity. The plot of ln q_e versus ln C_e had a slope with the value of $1/n$ and an intercept magnitude of ln K_F.

2.2.4. Fourier transform infrared spectroscopy (FTIR) analysis. FTIR analysis in the solid phase was performed using a FTIR spectrometer (Perkin Elmer). Adsorbent powders were blended with IR-grade KBr in an agate mortar and pressed into tablets. Spectra of the CMC-g-PAA hydrogel before and after Ni(II) sorption, and after desorption were recorded between 4000 and 400 cm^{-1}.

3. Results and Discussion

3.1. Adsorption isotherm

To investigate the adsorption capacity as a function of aqueous concentration of the heavy metals, adsorption isotherms are widely employed for fitting the data using representation of the adsorption (expressed as mg metal adsorbed per g of solid adsorbent) versus concentration of the heavy metals at equilibrium C_e. Adsorption isotherm for Ni^{2+} ions on different CMC-g-PAA hydrogel are shown in Figure 1.

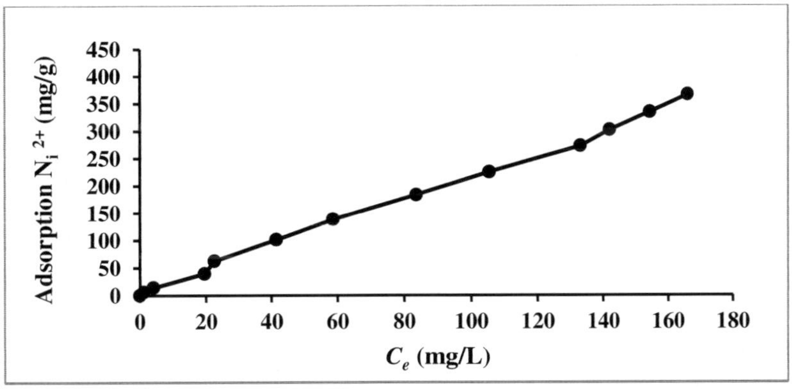

Figure 1. Adsorption isotherms for Ni^{2+} ions on CMC-g-PAA hydrogel.

In batch adsorption processes, the initial metal ion concentration provided an important driving force to overcome the mass transfer resistance of Ni(II) ions between the aqueous solution and the CMC-g-PAA hydrogel surface. As a result, the amount of Ni(II) ions adsorbed was expected to be higher with a higher initial metal ions concentration, hence improving the adsorption process [14]. From Figure 1, we found that when the initial Ni(II) ion concentration increased from 4.18 to 166.11 mg/L, the uptake capacity of the CMC-g-PAA hydrogel for Ni(II) ions increased from 14.18 to 366.11 mg/g. This could be explained from two aspects. Firstly, the higher initial Ni(II) ion concentration increased the driving force to overcome the mass transfer resistance of Ni(II) ions between the aqueous and solid phases resulting in a higher probability of collision between the Ni(II) ions and the CMC-g-PAA hydrogel. Secondly, the increase of Ni(II) ion uptake capacity of the CMC-g-PAA hydrogel for Ni(II) ions with increasing initial Ni(II) ion concentration might also be attributed to a more intense interaction between the Ni(II) ions and the CMC-g-PAA hydrogel [14].

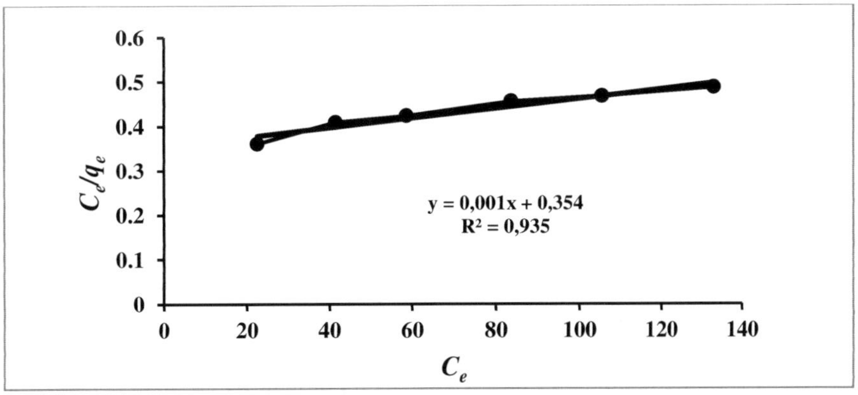

Figure 2. The Langmuir adsorption isotherm for Ni^{2+} ion adsorption by CMC-g-PAA hydrogel (adsorbent dosage 100 mg/L; pH 5; initial metal concentration 10 mg/L; 30°C).

3.2. Langmuir model

The Langmuir equation has been frequently used to give the sorption equilibrium. The Langmuir model assumes that the uptake of metal ions occurs on a homogenous surface by monolayer adsorption without any interaction between the adsorbed ions. All surface sites are alike and can only accommodate one adsorbed molecule, the ability of a molecule to be adsorbed on a given site is independent of its neighboring sites occupancy, adsorption is reversible, and the adsorbed molecule cannot migrate across the surface or interact with neighboring molecules [9]. To get the equilibrium data, initial metal concentrations were varied while the adsorbent mass in each sample was kept constant [11]. The linearized Langmuir isotherm allows the calculation of adsorption capacities and Langmuir constant by Equation (3).

The adsorption constants of Langmuir isotherm model parameters, q_{max} and K_L, were estimated from the intercept and slope of C_e/q_e versus C_e according to Equation (3) and obtained as 1000 mg/g related to adsorption capacity and 0.003 1/mg related to the rate of adsorption, respectively. The correlation coefficient of Langmuir isotherm (R^2) was 0.935 (Figure 2), which shows correlation or linear relationship. The relationship becomes more linear with R^2 values closer to 1. Thus, it was found that adsorption of Ni(II) onto CMC-g-PAA correlated well with Langmuir equation [1,11].

3.3. Freundlich model

The Freundlich isotherm was originally of an empirical nature, but was later interpreted as sorption to heterogeneous surfaces or surface supporting sites of varied affinities. It is assumed that the stronger binding sites are occupied first and that binding strength decreases with increasing degree of site occupation. The Freundlich isotherm can describe the adsorption of organic and inorganic compound on a wide variety of adsorbents [9].

On average, a favorable adsorption tends to have Freundlich constant n between 1 and 10. Larger value of n (smaller value of $1/n$) implies a stronger interaction between the adsorbent and the adsorbate while $1/n$ equal 1 indicates linear adsorption leading to identical adsorption energies for all sites. Linear adsorption generally occurs at a very low solute concentration and low loading of the adsorbent [9].

The value of n of this model fell in the range of 1–10, indicating favorable biosorption. The numerical value of $1/n < 1$ indicated that adsorption capacity was only slightly suppressed at lower equilibrium concentration. This isotherm did not predict any saturation of the adsorbent by the adsorbate. Thus infinite surface coverage was predicted mathematically, indicating multilayer adsorption on the surface. The present study results indicated that the Freundlich model fit the experimental data well.

In this study Freundlich model was the best-fit isotherm for adsorption of Ni(II) to CMC-g-PAA hydrogel. Freundlich isotherm model parameters K_F and n, calculated according to Equation (5), were 4.614 for adsorption capacity and 1.2 for intensity of the adsorption. The correlation coefficient of Freundlich isotherm (R^2) was 0.999 (Figure 3).

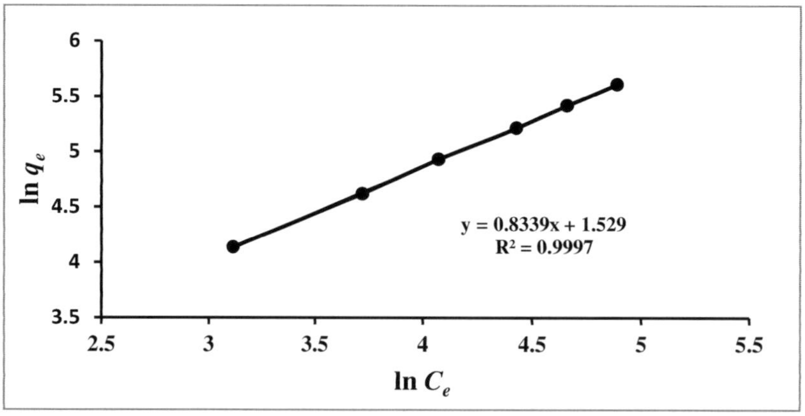

Figure 3. The Freundlich adsorption isotherm for Ni^{2+} ion adsorption by CMC-g-PAA hydrogel (adsorbent dosage 100 mg/L; pH 5; initial metal concentration 10 mg/L; 30°C.

Table 1. Parameter for the Langmuir and Freundlich adsorption isotherms for Ni^{2+} on the CMC-g-PAA hydrogel adsorbent.

Metal	Adsorbent dosage (CMC-g-PAA) hydrogel adsorbent (g/L)	Langmuir isotherm constants			Freundlich isotherm constants		
		K_L (1/mg)	q_{max} (mg/g)	R^2	K_F (1/g)	n	R^2
Ni(II)	1	0.003	1000	0.935	4.614	1.2	0.999

The characteristics of Langmuir and Freundlich equations are shown in Table 1. In Table 1, Freundlich model seems to fit the experimental data well. It has been stated that magnitude of the exponent $1/n$ gives an indication of the favorability or adsorption capacity of the solid adsorbent for the studied adsorbate. Values $n > 1$ represented favorable adsorption conditions [1,15].

3.4. FTIR spectrum

Figure 4 shows the FTIR spectra of acrylic acid, the CMC-g-PAA hydrogel before adsorption, hydrogel after adsorption, and hydrogel after desorption. For FTIR spectra of acrylic acid (a) and cellulose adsorbent (b), there was a small shift of the ring vibration peak at 1712 cm^{-1}, 1626 cm^{-1}, 1425 cm^{-1}, 1299 cm^{-1}, 1240 cm^{-1}, 1060 cm^{-1}, and 983 cm^{-1} in acrylic acid to 1708 cm^{-1}, 1550 cm^{-1}, 1434 cm^{-1}, 1224 cm^{-1}, 1163 cm^{-1}, 1068 cm^{-1}, and 1002 cm^{-1} for carboxymethyl cellulose, which suggested a successful grafting. A new band at 1550 cm^{-1} was related to the stretching vibration of –C=O. The peaks at 1434 cm^{-1} and 1163 cm^{-1} in the spectrum of CMC-AA (acrylic acid) adsorbent were related to the stretching vibration and symmetrical stretching of –COO groups [19].

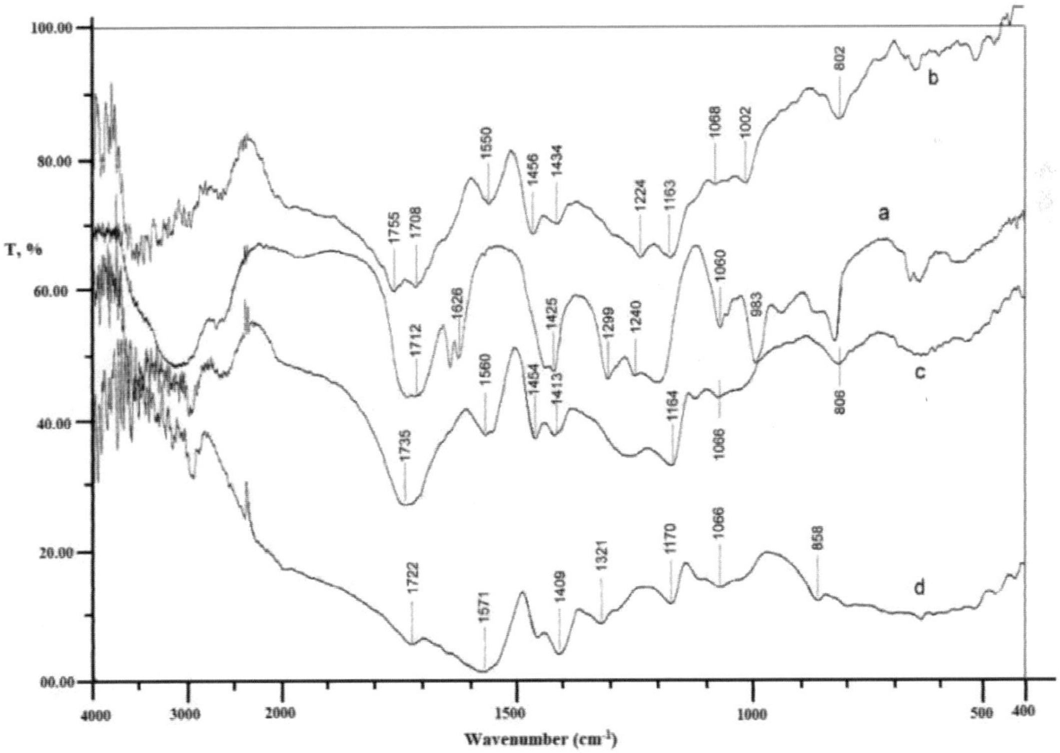

Figure 4. IR spectra hydrogel adsorbent: (a) acrylic acid; (b) CMC-g-PAA hydrogel before adsorption; (c) hydrogel after adsorption; (d) hydrogel after desorption.

To identify the possible functional groups on cellulose adsorbent involved in the binding of Ni(II), FTIR spectra were obtained before (b) and after (c) metal adsorption. If a ligand coordinated to a metal, the energy of the ligand material would most likely be perturbed, leading ultimately to subtle shifts in the absorption peaks in the FTIR spectra. These FTIR absorption bands were usually shifted to lower or higher frequencies. There was a small shift of the ring vibration peak at 1755 cm^{-1}, 1550 cm^{-1}, 1456 cm^{-1}, 1434 cm^{-1}, 1163 cm^{-1}, 1002 cm^{-1}, and 802 cm^{-1} in carboxymethyl cellulose adsorbent to 1735 cm^{-1}, 1560 cm^{-1}, 1454 cm^{-1}, 1413 cm^{-1}, 1164 cm^{-1}, 1066 cm^{-1}, and 806 cm^{-1} for Ni(II) loaded adsorbent. There were new bands at 1735 cm^{-1} (C=O stretching), 1454 cm^{-1} (CH_2 bending) and 1164 cm^{-1} in the spectrum of CMC-g-PAA adsorbent that were related to the stretching vibration and symmetrical stretching of –COO group [20].

3.5. SEM image

SEM images of carboxymethyl cellulose adsorbent before adsorption, after adsorption and after desorption are shown in Figure 5.

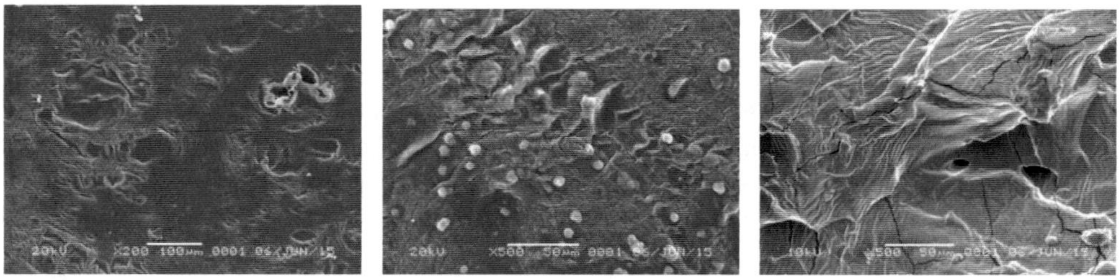

(a) (b) (c)

Figure 5. SEM micrograph of CMC-g-PAA hydrogel adsorbent: (a) before adsorption; (b) after adsorption; (c) after desorption.

The SEM micrograph of hydrogel adsorbent in Figure 5 (a) showed fractured and rough surface morphology that provided better adsorption sites. It can be observed that the surface of the adsorbent after adsorption (b) changed, and their visible pores of the adsorbent had disappeared or had been filled up. Also, it can be seen that the surface of adsorbent was back to become fractured and rough again after desorption process in three times [20].

4. Conclusion

The Freundlich isotherm model was represented by K_F and n empirical constants as characteristics of the system; K_F represent the adsorption capacity and n was indicating the intensity of adsorption. The Freundlich model was the best-fit isotherm for adsorption of Ni(II) to CMC-g-PAA hydrogel adsorbent with the correlation coefficient of Freundlich isotherm (R^2) was 0.999 and the magnitude of the exponent $1/n$ gave an indication of the favorability or adsorption capacity of the solid adsorbent for the Ni(II) adsorbate. Values $n > 1$ represented favorable adsorption conditions.

The Langmuir isotherm model indicated the intercept $1/(q_{max}.K_L)$ and slope $(1/q_{max})$ of C_e/q_e versus C_e. The adsorption constant of Langmuir isotherm model parameters are q_{max} and K_L. q_{max} is related to adsorption capacity and K_L is related to rate of adsorption. The Langmuir isotherm was a good fitting model for CMC-g-PAA hydrogel adsorbent with correlation coefficient (R^2) of 0.935.

Acknowledgement

The authors gratefully acknowledge the Research Unit for Clean Technology LIPI, Bandung, Indonesia, for providing financial support and facilities required to carry out this work.

References

[1] Nita I, Iorgulescu M, Spiroiu M F, Ghiurea M, Petcu C and Cinteză O 2007 *Analele Universitatii din Bucuresti-Chimie* **1** 59-67
[2] Ewecharoen A, Thiravetyan P, Wendel E and Bertagnolli H 2009 *J. Hazard. Mat.* **171** 335-9
[3] Ozay O, Ekici S, Baran Y, Aktas N and Sahiner N 2009 *Water Res.* **43** 4403-11
[4] Zheng Y, Hua S and Wang A 2010 *Desalination* **263,** 170-5
[5] Ekebafe L O, Ogbeifun D E and Okieimen F E 2012 *Afr. J. Env. Sci. Tech.* **6** 275-82
[6] Al-Abachi M Q, Al-Awady N S and Al-Anbakey A M 2013 *J. Al-Nahrain Univ.* **16** 30-9
[7] Salim A J 2015 *J. Al-Nahrain Univ.* **18** 40-8
[8] Nada A A M A, Alkady M Y and Fekry H M 2007 *BioResources* **3** 46-59
[9] Govindarajan C, Ramasubramaniam S, Gomathi T and Sudha P N 2011 *Arch. Appl. Sci. Res.* **3** 572-80
[10] Kong W, Ren J, Wang S and Chen Q 2014 *BioRes.* **9** 3184-96
[11] Shanmugapriya A, Ramya R, Ramasubramaniam S and Sudha P N 2011 *Arch. Appl. Sci. Res.* **3** 424-35
[12] Demirbas A 2008 *J. Hazard. Mat.* **157** 220-9
[13] Anah L, Astrini N and Haryono A 2015 *Macromol. Symp.* **353** 178-84
[14] Chen J H, Xing H T, Guo H X, Li G P, Weng W and Hu S R 2013 *J. Hazard. Mat.* **248-249** 285-94
[15] Gad H M H and Nabila A M 2014 *Int. J. Advanced. Sci. Technol. Res.* **4** 184-201
[16] Aksu Z 2001 *Sep. Purif. Technol.* **21** 285-94
[17] Ozsoy H D and Kumbur H 2006 *J. Hazard. Mat.* **136** 911-6
[18] Treyball R E 1988 *Mass Transfer Operations 10th ed.* (New York: McGraw-Hill)
[19] Zhang G, Yi L, Deng H and Sun P 2014 *J. Env. Sci.* **26** 1203-11
[20] Lin C, Qiao S, Luo W, Liu Y, Liu D, Li X and Liu M 2014 *BioResources* **9** 6998-7017

Effect of natural zeolite and kaolin as a catalyst in the isothermal-catalytic cracking of real municipal solid waste (MSW) for bio-oil production

I M Gandidi[1]*, M D Susila[2] and H Rustamaji[3]

[1] Department of Mechanical Engineering Education, Indonesia University of Education, Jl. Dr. Setiabudhi No. 229, Kota Bandung 40154, West Java, Indonesia
[2] Department of Mechanical Engineering, Faculty of Engineering, University of Lampung, Jl. S. Brojonegoro, Bandarlampung 35145, Indonesia
[3] Department of Chemical Engineering, Faculty of Engineering, University of Lampung, Jl. S. Brojonegoro, Bandarlampung 35145, Indonesia

*Email: indra.gandidi@upi.edu

Abstract. A real MSW conversion into bio-oil by pyrolysis method had been investigated. The pyrolysis experiment was carried out in an isothermal fixed bed reactor at 400 °C for 60 min. Natural activated zeolite and kaolin catalysts were employed in this process with catalyst-MSW at ratio of 0.5 w/w. Influence of activated catalyst to pyrolysis productivity and bio-oil properties were studied. Comparison between zeolite and kaolin catalytic pyrolysis was examined and bio-oil products were analyzed by GC-MS. The yields of liquid fuel or bio-oil were found around 23.6 and 21.4% for kaolin and zeolite catalytic pyrolysis, respectively. The carbon range distribution in bio-oil were 65.38% of gasoline range (C_5-C_{12}) and 33.49% of diesel range (C_{13}-C_{20}) for activated kaolin catalytic cracking. Activated zeolite had 51.62% of gasoline range (C_5-C_{12}) and 48.37% of diesel range (C_{13}-C_{20}). The results showed that both of bio-oil products mostly contained paraffin and olefin. The use of kaolin as catalyst was better than zeolite in converting the real MSW into bio-oil.

1. Introduction

The open dumping system of municipal solid waste (MSW) in Indonesia has created a lot of impacts for people and environment in which the system is no longer relevant due to the limitations in the urban area [1]. The landfill becomes a habitat for insects, pests and bacteria that can spread various diseases in the open environment. The piles of garbage in the landfill will produce some leachate that contaminates the ground and surface water [2]. Furthermore, MSW will be degraded biochemically and produce methane gas into the atmosphere. It is known that the potential of methane gas pollution is about twenty times more dangerous than the pollution by CO_2 [3]. In addition, MSW is one of major contributors for flood and increased risk of global warming and greenhouse gas effect [4].

On the other hand, MSW is a material contains energy in the form of chemical bonds between molecules of carbon, hydrogen and oxygen. When the chemical bonds are destroyed, the organic material will release the chemical energy in the form of gas, liquid and solid which is commonly called biofuel [5,6]. The calorific value content is around 20.57 MJ/kg [7]. Thus, MSW is one of the potential energy materials that can be converted into bio-energy. Pyrolysis is one method proposed for bio-energy production from MSW. Pyrolysis is a thermal degradation in which MSW is converted into solid fraction, liquid fraction consisting of paraffin, olefins, naphthene and aromatics, and gas at elevated temperature in the absence of oxygen [8]. Usually the temperature used is in the range of 400-600 °C. This process is intended to break down the long hydrocarbon chains into short-chain hydrocarbons. Hence, MSW appears to be eligible to be developed as a raw material for bio-oil production by pyrolysis technique [9-12]. MSW converted into bio-oil has attracted much attention due to its ease in storage, transport, and upgrading.

Recently, researches are focused on the catalysts to reduce the activation energy and to improve the thermal efficiency due to endothermic reaction at high temperature as well as the decomposition process improvement during catalytic cracking. The presence of catalyst will generate secondary cracking to the solids and liquids. Thus, resulted in higher pyrolytic products. Chika *et al* [13] carried out the catalytic pyrolysis by using Y-zeolite and ZSM-5 as catalyst to convert electric and electronic plastic waste into bio-oil, while Aida *et al* [14] used fluidized catalytic cracking (FCC) catalyst to convert plastic waste. Norbert *et al* [15] performed catalytic pyrolysis by Y-zeolite, FCC, MoO_3 and HZSM-5 to convert MSW and municipal plastic waste (MPW). Almeida and Marque [16] compared between thermal and HZSM-5-catalytic cracking to MPW. Funda *et al* [17] studied the catalyst efficiency of Y-zeolite, β-zeolite, equilibrium FCC, MoO_3, Ni–Mo-catalyst, and HZSM-5 on the catalytic cracking of MSW and MPW. The studies mentioned above, used synthetic or commercial catalysts in the pyrolysis process which is high costs of production, faster deactivation time of catalyst by MSW impurities or contaminant, as well as unfavorable in the viewpoint of economic issues.

Nevertheless, available natural catalysts as zeolite, kaolin, and dolomite have been investigated as catalysts on pyrolysis process instead of commercial catalysts. Wenger [18] performed the experiment of catalytic pyrolysis to produce bio-oil from mixed plastic waste by using clay as the catalyst. Kyaw *et al* [19] carried out pyrolysis to compare between clay and dolomite as the catalyst. Panda *et al* [20] used kaolin in the polypropylene pyrolysis and Mochamad *et al* [21] carried out two stage pyrolysis of HDPE and polypropylene as raw material with natural zeolite as the catalyst to produce bio-oil. However, single component pyrolysis or MSW that was specified such as pure and mixed plastic material for bio-oil production was performed. However, the MSW characteristic in developing countries contains a huge variety of mixed waste and it is impossible to collect different kinds separately because of the large quantities. In addition, more information due to bio-oil characteristics from the real unseparated MSW pyrolysis is necessary prior to use as an engine fuel. Moreover, natural catalysts have different characteristics from each other and it is very dependent on the mining location and chemical composition of the natural catalyst. For instance, Si/Al ratios in the catalyst shows strong effects on the pyrolytic products. Significant effect on the decomposition process will be obtained under lower Si/Al ratio and increases the acidity of catalyst [22]. The selection of appropriate natural catalyst for real MSW pyrolysis is required. In addition, kaolin and zeolite are not used as catalyst in the pyrolysis process of real MSW.

This research was focused on the productivity and selectivity of natural zeolite and kaolin as a catalyst on pyrolysis process to produce bio-oil from the real MSW. The real MSW consists of biomass waste, various plastics waste, textile waste, papers waste, and rubber waste.

The characterization was carried on the composition and yield, chemical, and physical properties of bio-oil. Comparison between pyrolysis and conventional oils such as diesel-48 and gasoline-88 was performed to examine the possibility for conventional fuels substitution.

2. Material and Methods

2.1. Waste and catalyst for pyrolysis

The MSW was collected from the final disposal site located in Bakung District, Bandarlampung City. This MSW consisted of various wastes such as vegetable waste from the traditional market, garden waste, household waste, plastic waste, paper/cardboard, rubber waste, and textile waste. Samples were sun dried for 3-5 days. The dried MSW was then chopped into pieces smaller than 5-10 cm to provide proportional size with reactor capacity. The dry basis composition of MSW are shown in Figure 1.

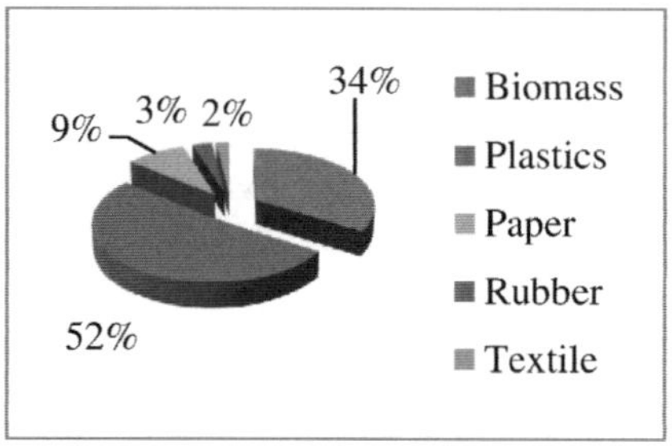

Figure 1. The composition of MSW.

The natural zeolite and kaolin used in this study were obtained from the local source in Lampung Province, Indonesia and then ground by a hummer mill and sieved to 20-30 mesh. Zeolite and kaolin were activated thermally at 500 °C for 2 h to improve the active site and crystallinity. To examine the structure and surface of the natural zeolite and kaolin, the XRD was employed. The composition of the activated natural zeolite and kaolin is shown in Table 1. The high amount of SiO_2 and Al_2O_3 in the natural zeolite and kaolin showed that zeolite and kaolin are proper to use as a catalyst. These compounds could increase the cracking efficiency and increase the pyrolytic products.

Table 1. X-Ray analysis for chemical composition of activated zeolite and kaolin.

Kind of catalyst	Compound	Concentration (%)
Zeolite	SiO_2	68.5
	Al_2O_3	13.17
	Fe_2O_3	2.98
	MgO	1.15
	CaO	2.47
	K_2O	1.80
	Na_2O	1.06
	TiO_2	0.14
Kaolin	SiO_2	43.12
	Al_2O_3	46.07
	Fe_2O_3	0
	MgO	0.27
	CaO	0.3
	K_2O	0.01
	ZnO	0.0064
	TiO_2	0.74

Source: CV. Mina Tama, Bandarlampung

Figure 2 shows the X-ray patterns for both catalysts. The sharp peaks implied that natural zeolite and kaolin had good crystallinity and possibly gave strong effects in the decomposition of MSW and stable at high temperature.

2.2. Pyrolysis experiment and bio-oil analysis

The MSW sample was pyrolysed using a fixed bed vacuum reactor equipped with air lock valve between reactor and hopper to avoid the air took place of reactor entry. The reactor was made from stainless cylinder with 310 mm in height and 160 mm in internal diameter. The electrical heating jacket was connected to thermo-controller and thermocouple was taped around the reactor. The reactor was set vertically and connected to a nitrogen tank using rubber tube and then N_2 gas was introduced into the reactor for 3 min from the top, passed through and flowed out through the top of the reactor. The flow of nitrogen replaced the air from the reactor and permitted the pyrolysis reaction took place under inert vacuum condition.

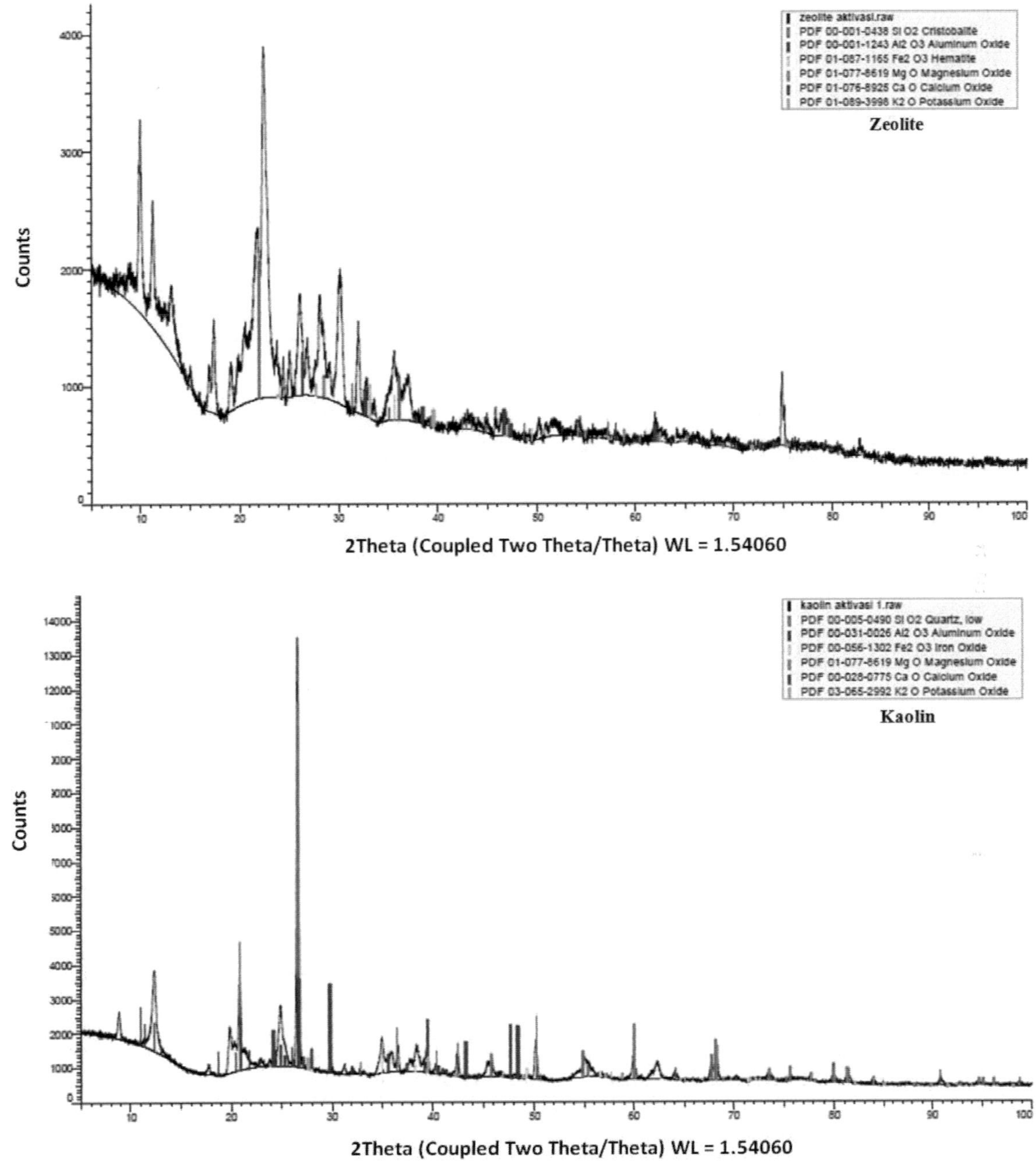

Figure 2. X-Ray patterns of the activated zeolite and kaolin.

Nitrogen gas was introduced into the reactor and then the reactor was heated until reactor temperature reached to 400 °C with heating rate of 12°C. An amount of 750 g of mixed catalyst (250 g) –MSW (500 g) was put into reactor by the opened air lock valve. The reaction time was set for 60 min for both of pyrolysis experiments. The gas produced was flowed into vacuum gas cleaner as the first condenser to the shell and tube condensers at 20°C as the secondary condenser to separate the permanent gas and liquid. The gas was burned off to prevent emission from hydrocarbon gases. The char remained in the reactor was collected after the pyrolysis reaction was finished. The experimental apparatus is shown in Figure 3.

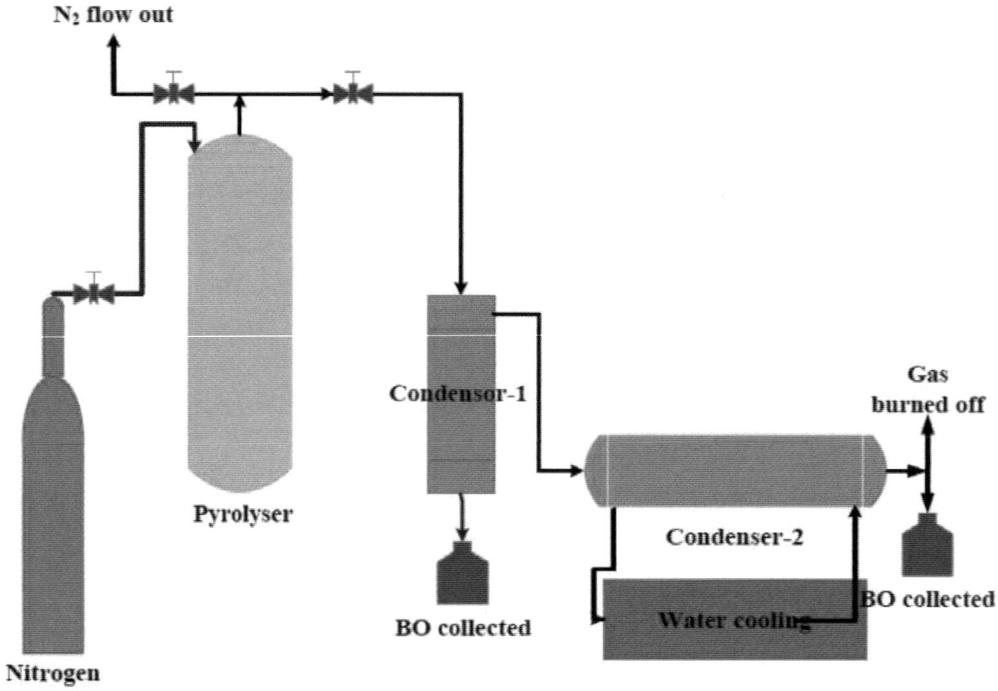

Figure 3. The experiment apparatus.

Regarding the pyrolytic products, products obtained from the pyrolysis could be divided into non-condensable gas fraction, solid fraction and liquid (bio-oil) fraction. From bio-oil yield that could be recovered, the hydrocarbon chains were in the gasoline range (C_5-C_{12}), diesel (C_{12}-C_{20}) and heavy oil (> C_{20}) that consisted of paraffin, olefins, naphthenes and aromatics [23]. The pyrolysis liquid was analyzed using gas chromatography-mass spectrometry (GC-MS, QP2010S Shimadzu) which could be used to identify carbon number range and chemical composition. The column was DB-1 (Agilent J 100% dimethyl polysiloxane) capillary column, 30 m length with 0.25 mm diameter and 25 μm film thickness. Helium was used as the gas carrier. The temperature was set in initial temperature of 60°C for 5 min followed by a heating rate of 5°C/min to 280 °C and held for 51 min. The estimation of the heating value of bio-oil was approached by carbon-hydrogen weight ratio [24]. Paraffin had the highest heating value followed by iso-paraffin, olefin, nepthenes and aromatic (PIONA). Also, the distributions of PIONA in each range of hydrocarbon were calculated to determine the selectivity behavior of natural zeolite and kaolin catalysts in the pyrolysis process.

3. Results and Discussion

3.1. Physical properties of bio-oil

Table 2 shows the density and viscosity for conventional fuel and bio-oil. It shows that both of bio-oils from the natural catalytic cracking had similar density values and significantly different viscosity properties. The kinematic viscosity of natural kaolin was lower than the natural zeolite. It indicated that the natural kaolin had a good cracking reaction compared with the natural zeolite. However, both of bio-oils had properties close to diesel-48 and were acceptable to be considered as a future energy. Both of bio-oils had high solid and water contents due to impurities in the unwashed real MSW and direct catalytic as well. The solid content in the bio-oil was affected by ash from the biomass material.

Table 2. The physical properties of bio-oil.

Properties	Unit - Standard	Diesel	Gasoline	Kaolin	Zeolite
Cetane octane number		48 (*cetane*)	88 (*octane*)	-	-
Density@15°c	g/cm^3 ASTM D-1298	0.815-0.87	0.7-0.78	0.8038	0.8076
Kinematic viscosity@40°c	cSt ASTM D-445	2.0-5.0	0.4-0.8	2.0345	2.4775
Flash point	°C ASTM D-93	min 60	min 40	min 5	min 5
Pour point	°C	max 18	-	-	-
Water content	%Vol ASTM D-1744	max 0.05	-	3	2.5
Oil content	%Vol ASTM D-1744	-	-	86	90
Solid content	%Vol ASTM D-482	-	-	11	7.5

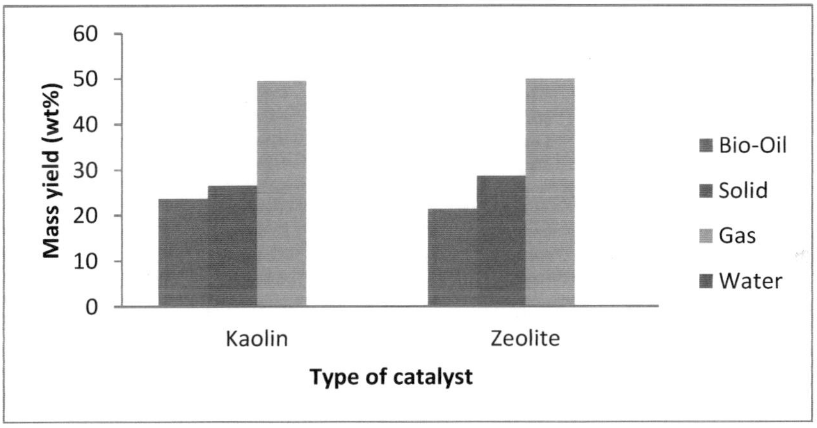

Figure 4. Mass yield of pyrolytic products.

3.2. Yield and chemical properties of bio-oil

The products from the MSW pyrolysis are shown in Figure 4. Catalytic cracking produced from natural kaolin and zeolite by pyrolysis is 23.6 and 21.4%, respectively. Kaolin shows higher yield compared to zeolite. This indicated that natural kaolin had a better productivity than the natural due its higher active site and higher acidity. However, the highest yield on the gaseous product and appearance of water content in the bio-oil implied that further drying for MSW was required and temperature of condensation environment around 20°C was unfavorable. Moreover, the low bio-oil yield could also be associated with the short reaction time and the long residence time of the gas in the reactor [25]. A similar result was presented by Panda and Singh [26] that used kaolin as a catalyst in plastic waste pyrolysis.

Figure 5 shows the carbon distribution range in the conventional and pyrolytic oils. The pyrolytic oil from mostly contains gasoline (C_5-C_{12}) followed by diesel (C_{13}-C_{20}) and a small amount of heavy oil (>C_{20}).

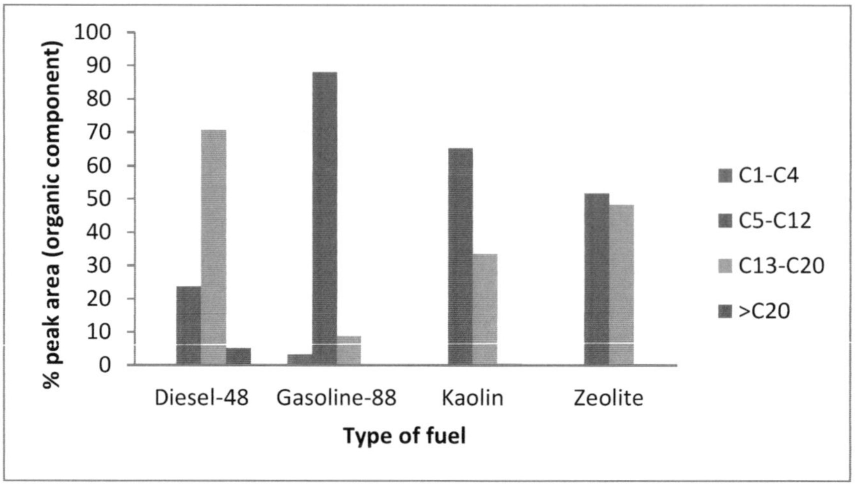

Figure 5. Carbon distribution range on conventional and pyrolysis fuel.

Bio-oil by natural kaolin catalytic consisted of 65.38% gasoline fraction and 33.49% of diesel fraction. Meanwhile bio-oil by natural zeolite showed 51.62% of gasoline fraction and 48.37% of diesel fraction. Clearly, natural kaolin had shown better properties as a catalyst for the production of bio-oil from MSW compared with natural zeolite, although natural zeolite did not have heavy oil fraction in the bio-oil.

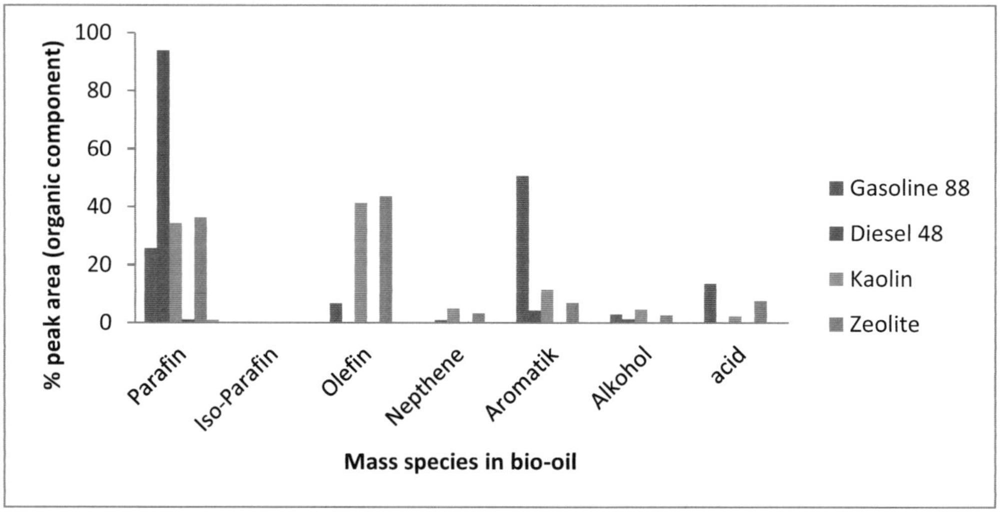

Figure 6. A comparison of mass species between bio-oil and conventional fuel.

Figure 6 reveals the organic components in the bio-oil in which most of organic components in the bio-oil were paraffin and olefin with small amount of aromatic, alcohol and acid. Only slight difference in the amount of paraffin and olefin between natural kaolin and natural zeolite in catalytic cracking process. Consequently, bio-oils from natural kaolin and natural zeolite catalytic pyrolysis showed the same caloric values. It illustrated that both of natural catalysts at 400 °C and 60 min had similarities in the selectivity behavior.

Figure 7 shows the PIONA contents in each hydrocarbon range on the pyrolytic bio-oil, reveals that gasoline fraction contained more olefin than paraffin, otherwise with the diesel fraction for both of pyrolytic bio-oils. This composition was significantly affected by the plastic and biomass in MSW which determined the chemical reaction during the process of pyrolysis. The presence of impurities and natural catalyst in the MSW was also allegedly plays an important role to the bio-oil composition [27].

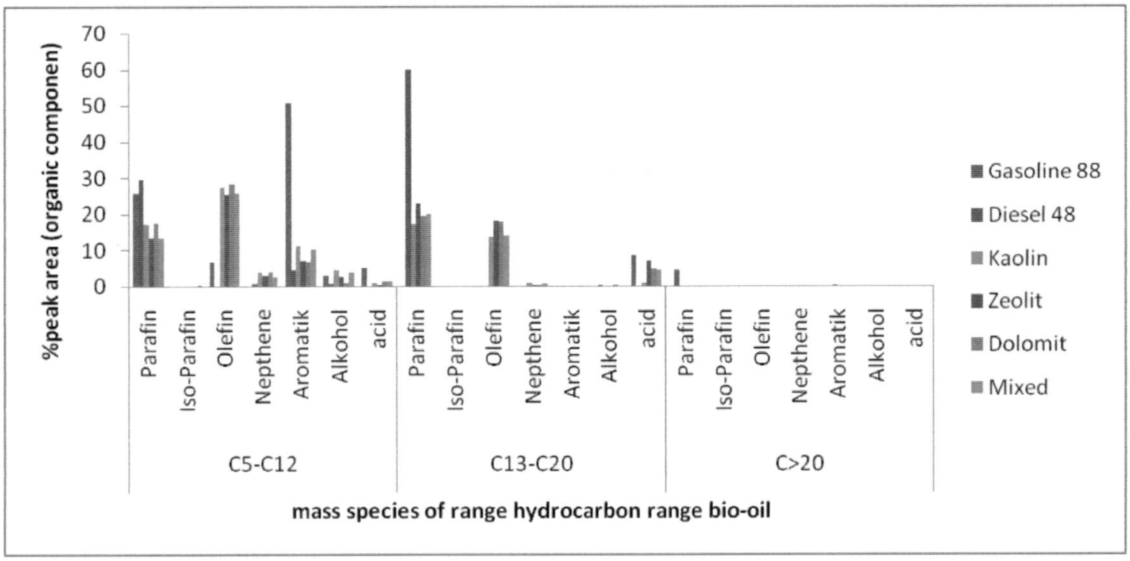

Figure 7. Number of mass species in each hydrocarbon range.

4. Conclusion

Real MSW is a great potential raw material to be converted into paraffin and olefin by thermal-catalytic cracking. Activated kaolin and zeolite can increase the productivity of bio-oil quite low in its selectivity Kaolin demonstrated a better capability for pyrolysis process than zeolite. The upgrading process to enhance the pyrolysis oil yield is necessary to obtain bio-oil with gasoline range.

References

[1] Arinal H, Gandidi I M and Harmen 2010 *Final Report of Strategic Research Grant*: *Integrated Waste Management System of Bandarlampung City as an Effort of Environmental Conservation and Bioenergy Production* (Bandarlampung, Indonesia: Lampung University)
[2] Le Courtouis A and World Bank 2012 *Municipal Solid Waste: Turning a Problem into Resource* http://www.ccacoalition.org/en/resources/municipal-solid-waste-turning-problem-resource
[3] Haradhan M 2012 **2** 3–10
[4] Hari B D and Sameer A 2013 *Habitat Int.* **38** 100–5.
[5] Ebru A and Ahmet D 2009 *5th Int. Advanced Technology Symp.*, Karabuk, Turkey
[6] McKendry P 2002 *Bioresource Technol.* **83** 37–46.
[7] Ari D P, Kilbergen W, Gultom and Aryadi S 2007 *Final Report of Feasibility Study: MSW-Electricity Power Plant in Bandung City* (Bandung, Indonesia: Bandung Institute of Technology)
[8] Hossain A K and Davies P A 2010 *Renew. Sust. Energy Rev.* **21** 165–89
[9] Liu Y Q, Wang J, Dong J and Thay J H 2012 3rd *Int. Conf. Industrial and Hazardous Waste Management* Singapore
[10] Seung S K, Jinsoon K, Jong K J, Young K P and Chan J P 2013 *Renewable Energy* **54** 241–7
[11] Velghe I, Carleer R, Yperman J and Schreurs S 2011 *J. Anal. App. Pyrolysis.* **92** 366–75
[12] Ayhan D 2004 *J. Anal. App. Pyrolysis* **72** 97–102
[13] Chika M, Jude A O and Paul T W 2015 *Catalytic Pyrolysis of Waste Plastic from Electrical and Electronic Equipment* (Leeds, United Kingdom: Energy Research Institute, Faculty of Engineering, University of Leeds)
[14] Aida Isma M I, Salmiaton A and Nur Dinie K B 2015 *Int. J. Env. Sci. Dev.* **6** 606–9
[15] Norbert M, Ates F and Borsodi N 2013 *Bioresource Technol.* **144** 370–9
[16] Almeida D and Marques M D F 2015 *Polimeros* **26** 44–51

IOP Conf. Series: Earth and Environmental Science **160** (2018) 012018 doi:10.1088/1755-1315/160/1/012018

[17] Funda A, Miskolczi N and Borsodi N 2013 *Bioresource Technol.* **133** 443–54
[18] Wenger J 2015 *J. Materials Science and Engineering* **5** 183–5
[19] Kyaw K T and Hmwe C S S 2015 *Int. J. Advances in Engineering & Technol.* **8** **(5)** 794–802
[20] Panda A K and Singh R K 2013 *Advances in Energy Engineering* **1** 74–84
[21] Mochamad S, Harwin S, Tinton N, Putri N, Shou C, Zainal A and Kunio Y 2014 *Energy Procedia* **47** 180–8
[22] Wang J L and Wang L L 2011 *Energy Sources.* **33** 1940–8
[23] Aguado J, Serrano D P, Guillermo S M and Madrid S 2007 *J. Anal. App. Pyrolysis* **79** 415–23
[24] Riazi M R 2005 *Characterization and Properties of Petroleum Fractions* (Philadelphia: ASTM International) pp 30–84
[25] Ayhan D 2009 *Energy Sources* **31** 1186–93
[26] Panda A K and Singh R K 2014 *Int. J. Environment and Waste Management* **13** 104–14
[27] Borsodi N, Miskolczi N, Angyal A, Bartha L, Kohan J and Lengyel A 2011 *45th International Petroleum Conference*, Bratislava, Slovak Republic

Economic situation of value chain actors in urban slums of Bandung: A case of Kiaracondong

M Ikemi[1*], K Ushijima[2], Y Otsuka[3], T Yamauchi[3], D Nilawati[4], D R Wulan[4] and N Sintawardani[4]

[1] Faculty of Tourism, Sapporo International University, Kiyota 4-1-4-1, Kiyotaku, Sapporo, Japan
[2] Northern Regional Building Research Institute, Hokkaido Research Organization, Midorigaoka Higashi 1-3-1-20, Asahikawa, Japan
[3] Graduate School of Health Sciences, Hokkaido University, Kita 12 Nishi 15, Kitaku, Sapporo, Japan
[4] Research Unit for Clean Technology, Indonesian Institute of Sciences (LIPI), Komplek LIPI Bandung, Jalan Sangkuriang, Gedung 50, Bandung 40135, Indonesia

[*]Email: mayu-ikemi@ts.siu.ac.jp

Abstract. This research project aims to develop a mechanism of sanitation value chain for people living in urban slums of Bandung City, Indonesia. These are generally low-income areas with poor hygienic environment. In order to grasp their actual living conditions, field surveys were conducted in August, September and November 2016. With regard to the findings of the surveys, household income and educational levels of people living in the urban slum are obviously low compared to the national average levels, especially in income. Among them, however, even women with lower academic backgrounds can support their families and maintain a certain level of living. In terms of the priority of spending money, it is placed on family healthcare and education for their children. As for borrowing and lending money, these activities are not common for the residents on a daily basis. On the other hand, when it comes to savings, they use not only banks but also local financial services managed by communities such as schools and cooperative societies. The results suggest that sanitation value chain involving economic activities should incorporate with family healthcare, child education and saving service in their local community.

1. Introduction

For many reasons, sanitation system is essential for human lives, especially in modern urban areas. It helps to promote public health, prevent environmental pollution, conserve ecosystem functions, be able to recycle resources, and achieve many other benefits. Due to the importance of sanitation system, there have been many studies reported on the system. Among the latest studies, Funamizu (2017) performed the analysis of value chain from people's sanitation. His analysis shows that activities of composting human feces, using urine as a liquid fertilizer and re-using treated gray water as irrigation water are to increase the cash income of people [1]. A study by Ushijima *et al* (2015) proposes a sanitation business model, which can create value chain of local residents and contribute to improving both their sanitation and income [2]. Scott *et al* (2017) identify the importance of opportunities to support activities of small and medium-sized enterprises in sanitation including innovative finance, training and business development assistance, and reinforcement of representative associations in low-income urban communities [3].

The purpose of our research project is to develop a mechanism of sanitation value chain for people living in urban slums of Bandung City, Indonesia. The slums in the city have been facing many difficulties, such as low-income and poor hygienic environment, and that is why the area was selected as the research site. On the basis of a number of previous studies, we define the concept of sanitation value chain as a chain of health, nutrition, food, income, and better life environment in this project entitled "The Sanitation Value Chain: Designing Sanitation Systems as Eco-Community Value System", which has been promoted by Research Institute for Humanity and Nature (RIHN). In order to carry out the research project successfully and sustainably there, the local residents' cooperation and participation are essential. They should be key actors to create and maintain value chains in their communities. Based on the aspects, it is necessary to grasp and understand their actual living conditions before planning to put the project into action. In this case study as a pre-research, we examine an effective means for the development of sanitation value chain in their community while taking advantage of their local characteristics.

2. Methodology

The research methods were questionnaire and interview targeting local residents living in low-income area. Focus group discussion was also utilized to understand better the residents' common values and opinions on their community life, because it is a useful research method to clarify and explore sensitive and personal issues. Due to the limited length of stay and schedule of the research team, we had to conduct the field surveys twice between August and November 2016 in an urban slum in Indonesia. The research area was Kecamatan Kiaracondong located in the central part of Bandung City. The target area was Rukun Warga No.2 (RW2) in Kelurahan Sukapura as shown in Figure 1. The population of RW2 area is 1,291 as of August 2016. This paper focuses on the survey results of local residents' economic situations mainly concerned with income level, spending and handling money, and the relationship with educational background. More details on the respondents of each field survey will hereinafter be described in section 3.1 and 3.2.

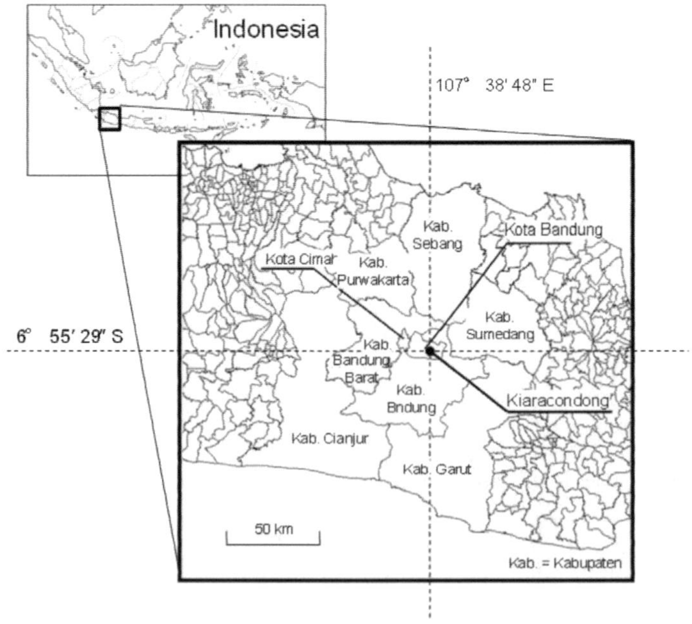

Figure 1. Research area.

3. Research results

3.1. Field survey with randomly selected men and women

One of the two field surveys was performed in November 2016. In this survey, questionnaire and interview with local residents were carried out. The respondents were over 20 years old who have a good understanding of their family situation, and they were randomly selected by the local research assistants who belong to a community-based youth organization (Karangtaruna). The total number of respondents was 234 that consisted of men and women with fifty-fifty ratio. The average age of respondents was 49.5 years (youngest: 22 years, oldest: 81 years). The questions of the survey were about household composition, income, expenditure, durable possessions, educational background, and so on.

With respect to the survey results on the respondents' educational status, there was no one who has no schooling. The pie chart in Figure 2 shows percentages by educational background in total respondents. The percentages of Primary (Sekolah Dasar: SD), Middle (Sekolah Menengah Pertama: SMP), Secondary (Sekolah Menengah Atas: SMA, and Sekolah Menengah Kejuruan: SMK), and Post-Secondary (Sarjana: S1, University, or Diploma) are 22%, 23%, 44%, and 11%, respectively. According to UNDP report (2016), Indonesia's gross enrolment ratios of Primary, Secondary, and Post-secondary were 106%, 82%, and 31%, respectively [4]. When we consider the UNDP data and the compulsory education system in Indonesia, it illustrates that the respondents' education levels are obviously lower than the national average.

Figure 3 shows the respondents' monthly household income. Most of them are less than 3,000,000 Indonesian Rupiah (IDR). Based on the data reported by World Bank on gross national income (GNI) per capita in 2016, the average monthly income per capita (not per household) in Indonesia can be calculated as approximately 3,538,000 IDR (about 0.00008 US dollars to the IDR) [5]. The respondents' average number of household members is 3.6 people, which is almost the same size as the national average of 3.9 people [6]. Taking into account this data, the survey result illustrates that the residents' incomes are significantly lower than the country average. This information will help us understand residents' economic burden of future project in their community.

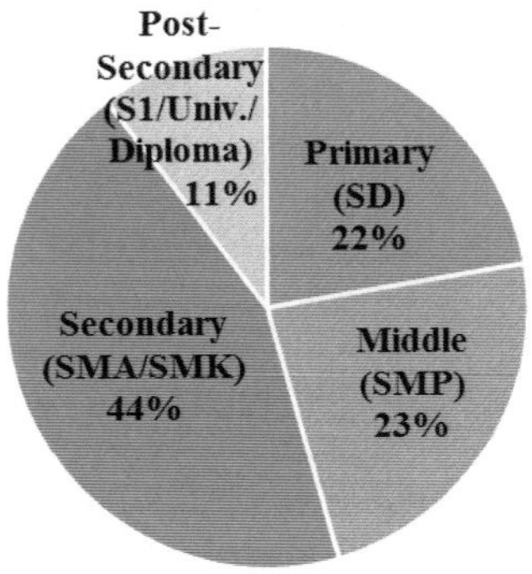

Figure 2. Educational background.

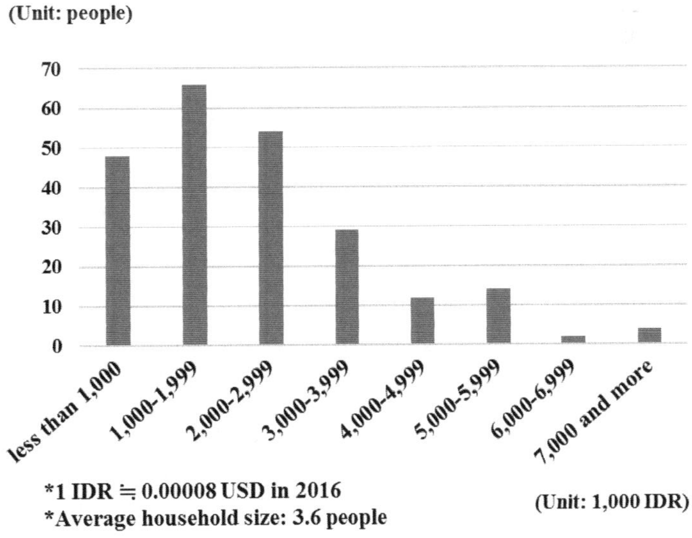

Figure 3. Monthly household income.

Figure 4 represents the relationships between income levels and educational backgrounds. The respondents' monthly household income levels are divided into 4 groups, "less than 2,000,000 IDR", "from 2,000,000 to 4,000,000", "from 4,000,000 to 6,000,000", and "6,000,000 or more". The results demonstrate that, the higher their income level is, the larger ratio of higher educational backgrounds. It means that higher educational group (Secondary or Post-Secondary) earns more income than lower educational group (Primary or Middle).

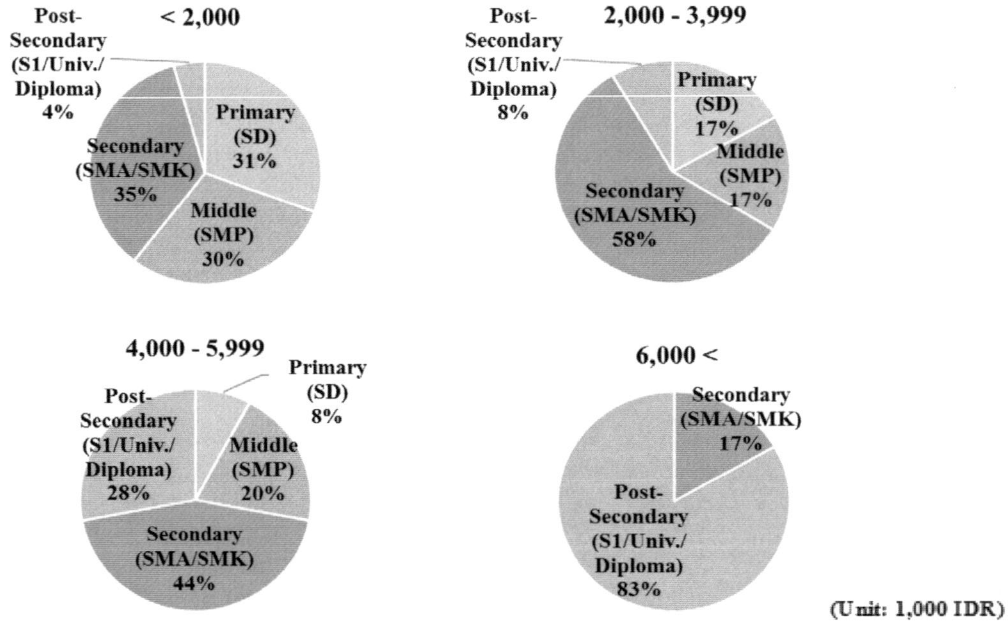

Figure 4. Composition of educational background par income level.

Figure 5 shows the average number of their durable possessions per household. TV and motorcycle are possessed at least one or more in each household. Refrigerator ownership is nearly one per household, however, laundry machine and PC are owned by nearly one in two households. As for car, only one in ten households possessed it.

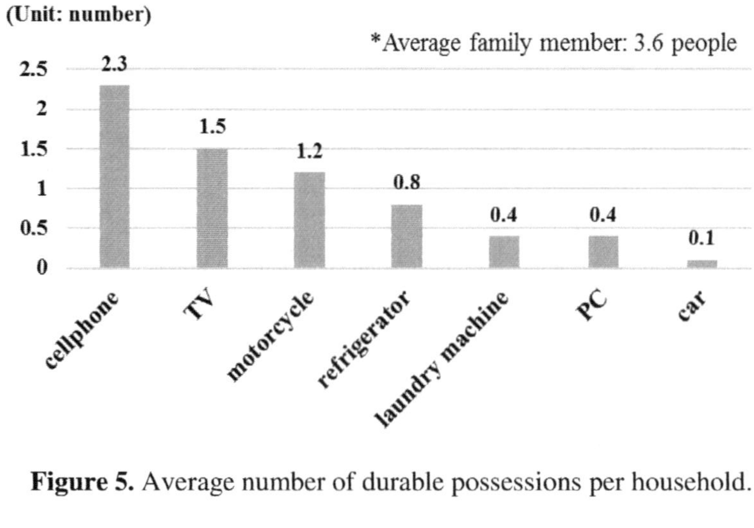

Figure 5. Average number of durable possessions per household.

From the aspect of water and sanitation, the situation of septic tank installation in their house was investigated. To the question "Does your house have a septic tank?", 60% (140 people) of total respondents said "yes", and 40% (94 people) said "no". To show the relationship between the level of income and possession of a septic tank, the percentages of respondents of each income level with and without a septic tank are displayed in Figure 6. When it comes to percentages of residents in the lowest income level (less than 2,000,000 IDR), the percentage among the residents who have a septic tank (45%) is less than the percentage among the residents who do not have a septic tank (57%). However, when focusing on relatively higher income levels (4,000,000 IDR or more), there is only 2% difference between those percentages of having a septic tank (15%) and having no septic tank (13%).

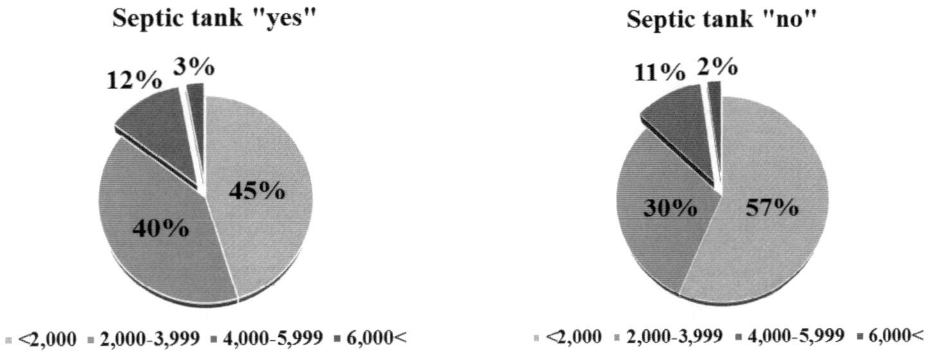

Figure 6. Comparison of income levels between having and not having a septic tank.

3.2. Field survey with married women

There was another field survey carried out in August and September 2016. It was conducted earlier than the survey introduced in Section 3.1., but these two surveys are mutually independent with different respondents. The methods in this survey were questionnaire, interview, and focus group discussion targeting married women. The total number of respondents was 30 and the average age was 44.4 years (youngest: 26, oldest: 67). The contents of the survey were household composition, income, expenditure, durable possessions, priority of spending money, borrowing and saving money, educational background, awareness and values for living conditions, and so on.

Most of the respondents are members of a women's community group called Pembinaan Kesejahteraan Keluarga (PKK) which means "family welfare movement". The PKK has been practiced to mobilize female participation on the national development policy since the late 1960's under the Soeharto administration [7]. The PKK members can be regarded as key residents who have leadership potential and ability to play an important role in sustainable development of urban slums in Indonesia. They are also expected to be main actors who create and expand value chains in their local communities through both sanitation improvement and income increase.

Regarding the composition ratio of respondents' educational backgrounds, Primary (SD) is 17%, Middle (SMP) is 30%, Secondary (SMA/SMK) is 33%, and Post-Secondary (S1/Univ./Diploma) is 20%. It is worthwhile to mention that the respondents are not over-concentrated within a group of educational background (highest: 33%, lowest: 17%). In other words, female residents living in the slum area have diverse educational achievement.

Table 1 shows the relationship between their educational backgrounds and monthly household income. The people with higher educational level (Secondary or Post-Secondary) have relatively higher income than lower educational level (Primary or Middle). It is similar to the result of previous survey discussed in section 3.1. However, some of them with lower educational level have even higher income than the people with higher academic background. It implicates that even women with lower academic backgrounds can somehow gain money (with financial support by their husband, parents, children, etc.) to maintain a certain level of living and support their family.

Table 1. Relationship between educational backgrounds and monthly household income.

(Unit: people)

(Unit: 1,000 IDR)	Primary	Middle	Secondary	Post-Secondary	Total
less than 1,000	1	0	0	0	1
1,000-1,499	0	4	0	0	4
1,500-1,999	0	0	2	0	2
2,000-2,499	2	1	3	1	7
2,500-2,999	1	0	1	1	3
3,000 and more	1	4	4	4	13
Total	5	9	10	6	30

Concerning the frequency of borrowing money, more than 80% (25 people) of the respondents answered "once a year or less". Among the rest of respondents, three women selected "never", one selected "almost every month", and the last woman indicated "no answer". The conducted interview revealed that there are few people who have the custom of borrowing or lending money in their living area. Especially on the activity of borrowing money, it is a common belief among interviewees that even if a person borrows money, it must be a secret. Figures 7 and 8 show the results of answers on the two questions: "From whom you borrow money?" and "For what you borrow money?". According to their responses, most people borrow money from a bank or cooperative society. This implies that even the residents living in urban slums can have access to formal financial services with relative ease. The most common purpose of borrowing money is for healthcare. Capital stock (such as entrepreneurship), education, and home renovation follow the healthcare for the purposes of borrowing money.

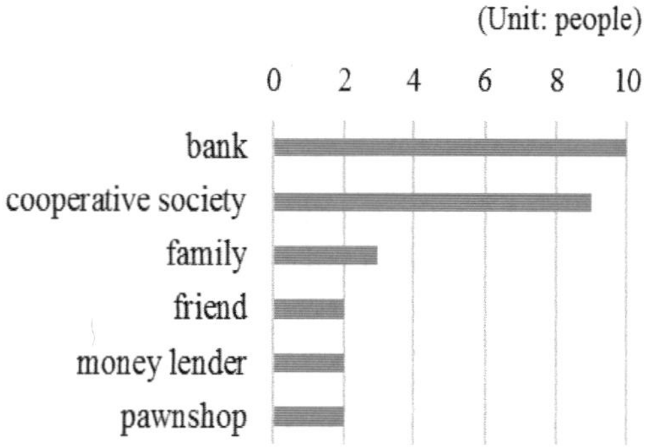

Figure 7. From whom do you borrow money?

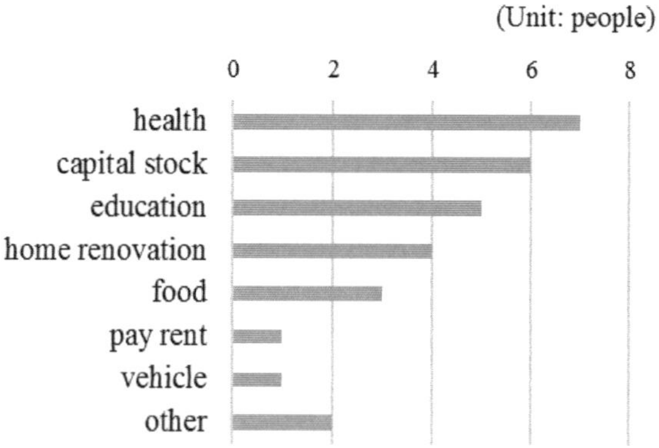

Figure 8. For what do you borrow money?

On the question of whether they save money, 9 out of the total 30 respondents answered "No, I don't.", and the rest of 21 answered "Yes, I do". Among the respondents who save money, half of them use a bank to store their money as shown in Figure 9. The other half of them use local financial services provided by communities and resident organizations other than banks such as schools, cooperative society, Rotating Savings and Credit Association (ROSCA), and so on.

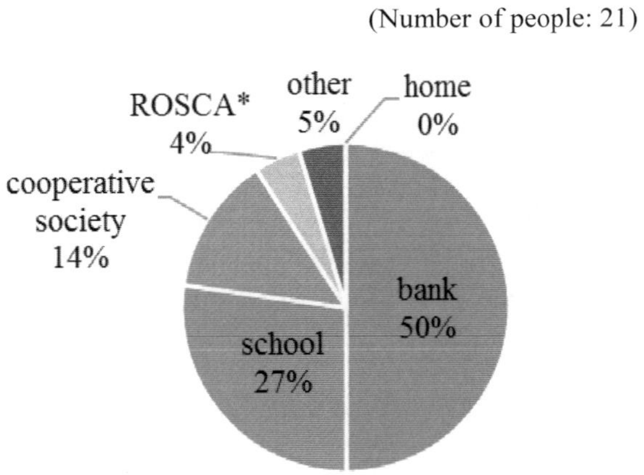

*ROSCA: Rotating Savings and Credit Association

Figure 9. Where do you save money?

With respect to expenditure, the average monthly expenditure per person is 461,000 IDR with the minimum of 25,000 IDR and the maximum of 1,500,000 IDR. Figure 10 shows the respondents' priority order on expenditure items. Almost all the respondents gave the first rank to food, but education was also perceived as of most importance by two people. As for the second rank, most of them selected healthcare, education and utilities as the next most important expenditure items.

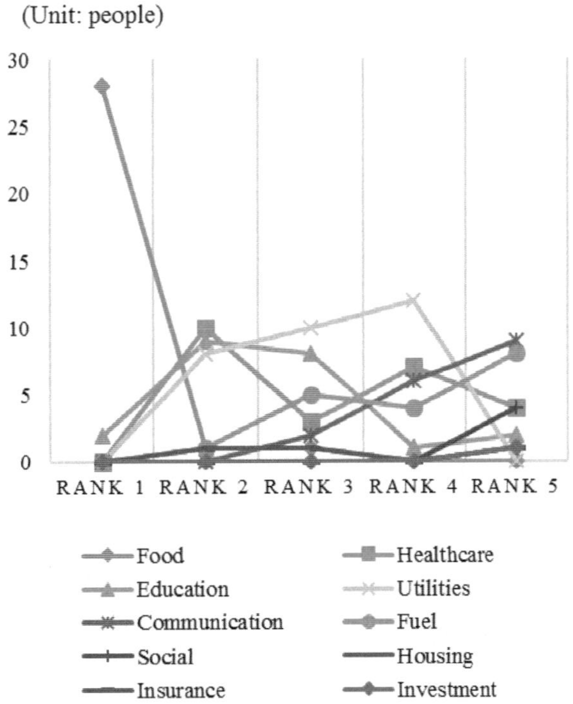

Figure 10. Importance of expenditure.

According to interview and focus group discussion surveys, food purchases accounts for the majority of their daily expenditures. They purchase daily meals and other grocery at a market or at a street stall (or kiosk) called "warung" which is a small shop generally managed in the owner's house. They go to a market about once a week on average, but they go to "warung" almost every day since the food is fresh, cheap, and nearby their home. They sometimes go to a supermarket in downtown about once a month to make a bulk purchase of groceries and food with a long shelf life. Nevertheless, they recognize that going to the market and supermarket leads to waste of money caused by impulse purchase. In contrast, the stall does not provide many choices. It normally provides the food items that residents most need to buy. Doing grocery shopping at a "warung" is more economical and better to save time. They also indicated that what to buy and where to buy would be decided by not only the price but also the food safety and sanitation.

Concerning positive aspects of living in their residential area, they said that "living costs are not high," and "location is convenient." They also indicated "easy access to hospital," "schools are within walking distances" and "kindergartens and universities are also close to the area." As for the negative aspects, most respondents indicated overpopulation in the area.

4. Discussion and summary

In this paper, we introduced and discussed economic situations of low-income people residing in an urban slum area of Bandung City, through the case study of Kiaracondong. In summary, the results of the first survey show that compared to the national average levels, income and educational levels of residents in the slum area are quite low, especially in income. Regarding their water and sanitation situation, among the residents who do not have a septic tank, more than half of them are in the lowest income level (less than 2,000,000 IDR). However, among the residents who have a septic tank, the ratio of residents in higher levels of income (4,000,000 IDR or more) is not that significantly different from that of among the residents who do not have a septic tank.

According to the results of the second survey, even women with lower academic backgrounds can support their families and maintain a certain level of living. The priority of spending money is placed on family healthcare and education for their children. Regarding borrowing and lending money, these activities are not common for the residents on a daily basis. Even if they borrow money occasionally, the main reasons are related to healthcare, education, and capital stock for small business. When it comes to savings rather than debts, they use not only banks but also local financial services managed by communities such as schools and cooperative societies.

In conclusion, we emphasize that it is crucial to consider local residents' needs for the development of sanitation value chain in their community, especially with their strong motivation and sustainable participation which no-one can tell unless the local residents show them by themselves. From the point of view, this study can suggest that sanitation value chain involving economic activities should be incorporated with family healthcare and child education for the participants because the results of our research show their high interest on them even if their economic condition is difficult.

Acknowledgments

This research was supported by "The Sanitation Value Chain: Designing Sanitation Systems as Eco-Community Value System" Project, Research Institute for Humanity and Nature (RIHN, Project No.14200107) headed by Professor Taro Yamauchi and Professor Naoyuki Funamizu. We would like to express our gratitude to all colleagues of the project and local residents in Kiaracondong for their great assistance and cooperation.

We would like to thank two anonymous reviewers for their time and effort in reviewing the papers and providing helpful and constructive comments.

References

[1] Funamizu N 2017 *Sanitation Value Chain* **1** 3-13
[2] Ushijima K, Funamizu N, Nabeshima T, Hijikata N, Ito R, Sou M, Maiga A H and Sintawardani

N 2015 *Water Policy* **17** 283-98
[3] Scott P, Forte J and Mazeau A 2017 *Barriers and Opportunities for Sanitation SMEs: A Study of the Wider Market System in Ghana* (London: Water & Sanitation for the Urban Poor (WSUP))
[4] United Nations Development Programme (UNDP) 2016 *Human Development Reports 2016 Human Development for Everyone* (New York: UNDP)
[5] World Bank 2016 *The World Bank Data: GNI per capita, Atlas method (current US$): Indonesia* https://data.worldbank.org/indicator/NY.GNP.PCAP.CD?locations=ID (Accessed on September 5, 2017)
[6] ArcGIS 2016 *Indonesia Average Household Size* (data source: Michael Bauer Research) http://www.arcgis.com/home/item.html?id=33ad793233cd47d885f8c126b51389af (Accessed on September 6, 2017)
[7] Hirano K 2005 *F-GENS Journal* **3** 261-8

Spatial planning for potential green TOD using suitability analysis at the metropolitan region scale

H M Taki[1,2*] and M M H Maatouk[1,3]

[1] Department of Urban and Regional Planning, King Abdulaziz University, Jeddah, KSA
[2] Department of Geography, University of Indonesia, Depok, Indonesia
[3] Department of Architecture, Minia University, Minia, Egypt.

[*]E-mail: htaki0001@stu.kau.edu.sa

Abstract. Public life is heavily influenced by the circulation and activity movements within the city. Transportation is one of the most influential factors in urban development. Transit Oriented Development (TOD) is one of the most innovative concepts in urban development related to transportation system. This paper aims to prescript potential areas for spatial planning of green TOD using a combined method of Geographic Information System (GIS) and Analytical Hierarchy Process (AHP) approaches to be implemented in Jakarta Metropolitan Region (JMR). Multi-indicator and weighting techniques were applied based on the perceived conclusions of the respective experts and professionals. Our analysis showed that the highly suitable class covered small area and TOD planning was focused in a specific area. The conclusion of this paper exhibited that there was a higher potential green TOD cluster occurred compared to its surrounding area. It meant that the ideally proportional balanced area for urban residential-commercial mixed use was in high preference of green TOD level.

1. Introduction

Each country has unique territorial characteristics which require different planning. Spatial planning is part of the urban development with the aim to creating liveable, efficient, and sustainable life resulting in a spatial plan of activities. Optimal spatial planning depends on the capacity of natural resources and human interaction as a user space. Therefore, the result is the creation of a harmonious relationship between space and its inhabitants [1–3]. Good spatial planning is planning that has several things, namely (1) Planning has a relation system between the function or area that can be accessed by all citizens, not separated; (2) Good relationship between building mass and open space to meet the standards of the ideal planning; (3) Planning has its purpose as an activity-generator that becomes the activity center of its citizen. The spatial plan requires the support of accurate spatial and non-spatial data in order to illustrate the real conditions of a targeted area [4]. Furthermore, it requires appropriate decision-making methods on complex issues to achieve the best-case TOD locations scenario and highest priority. In this case, Geographic Information System (GIS) and Analytic Hierarchy Process (AHP) can be utilized to answer the needs of spatial planning [4–8].

Geographic Information System (GIS) is a specific information system that manages various spatial data, non-spatial data, static data, and dynamic data (maps, satellite imagery, field surveys). The data processed in the GIS uses a database management system (DBMS) in which all types of data are stored in computer systems that could build, store, manage and display geographic information using software engine such as ArcInfo, MapInfo, or ArcView. Thus, it can build Decision Support System (DSS), especially for TOD location decisions [5]. AHP is a hierarchical structure method for solving complex and unstructured decision-making problems by giving subjective values of the relative importance of each indicator. It specifies which indicator has the highest priority to influence the outcome of the situation. AHP's key advantage is its ability to control the consistency of the decisions obtained to establish the best alternative from several options, based on criterion with specific weights [6].

The integration of combinations between GIS and AHP developed by [7–9] is applicable to the planning of Green Transit Oriented Development (TOD). The physical structure of the region should be supported by the transport system network. Transit, pedestrian, and cyclists should have broader access and mobility throughout the region [10]. In contrast, the reduction of the level of car use, would lead to a creation of safer areas; meaning less 'tailpipe emissions from vehicles', traffic accidents and crime rate.

1.1. Study area of research

The government of Indonesia has determined the Botabek area (supporting cities around Jakarta) as a buffer zone of Jakarta. Botabek have Spatial Law (RTRW). These RTRWs automatically contain designated zones allocated for trade and commerce activities (Jakarta's downtown); industrial development in Cibitung, Cikarang (Kab.Bekasi) and Cikupa (Kab.Tangerang); satellite cities in Bekasi City, Tangerang City, Serpong, South Tangerang City and Depok City and their supporting schools, shopping centers, hospitals and entertainment venues.

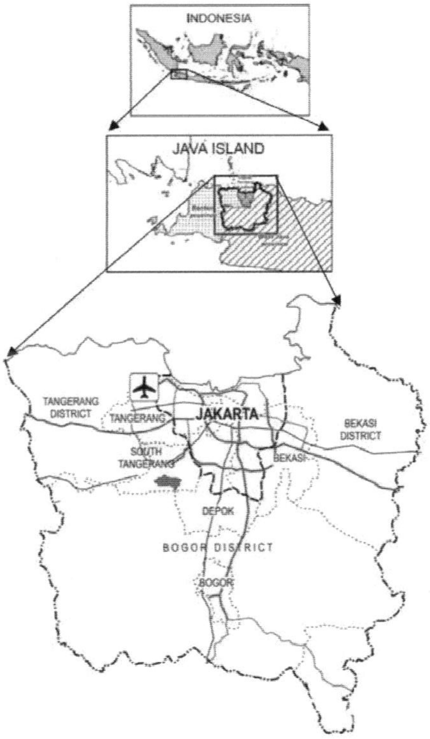

Figure 1. Study area of Jakarta Metropolitan Region (JMR).

2. Materials and Methods

2.1. Data collection

AHP primary data were done by interviewing 12 experts for getting their opinions. While the secondary data on spatial were collected from the institutions and agencies afterwards observation survey in the field.

Table 1. List of data layer for the TOD suitability and their original sources.

Data name	Type	Scale	Date	Source
Airport and Seaport	point	50.000	2010	Ministry of Transportation
Terminal	point	50.000	2010	Ministry of Transportation
Busway shelter	point	50.000	2010	Ministry of Transportation
Main road	line	50.000	2010	Ministry of Transportation
Tollgate	point	50.000	2010	Ministry of Transportation
Transportation facility	area	50.000	2014	Ministry of Agrarian and Spatial Planning
Education and public facility	area	50.000	2012	Ministry of Agrarian and Spatial Planning
Recreation facility	area	50.000	2012	Ministry of Agrarian and Spatial Planning
Hospital	point	50.000	2014	Geospatial Information Agency
School	point	50.000	2014	Geospatial Information Agency
BCR (building coverage ratio)	area	50.000	2015	Consultant project archive
Park	area	50.000	2012	Ministry of Agrarian and Spatial Planning
Planned housing	area	50.000	2012	Ministry of Agrarian and Spatial Planning
Residential	area	50.000	2012	Ministry of Agrarian and Spatial Planning
FAR (floor area ratio)	area	50.000	2015	Consultant project archive
Population	attribute	50.000	2016	Central Bureau of Statistics
CBD	point	50.000	2014	Geospatial Information Agency
Industrial	area	50.000	2012	Ministry of Agrarian and Spatial Planning
Commercial and business	area	50.000	2012	Ministry of Agrarian and Spatial Planning
Real estate	area	50.000	2012	Ministry of Agrarian and Spatial Planning

2.2. Criteria and sub-criteria

The explanations regarding these criteria and sub criteria were arranged in the hierarchy described in Figure 2 below.

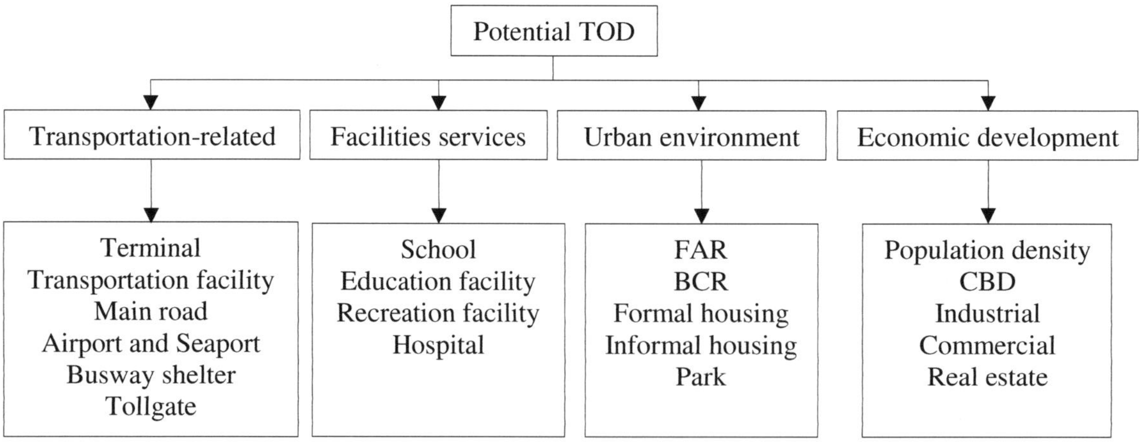

Figure 2. Criteria and sub criteria hierarchy.

2.3. Methodology

Methodology in this paper includes: (1) Determining the criteria and sub criteria of TOD from study literature; (2) Developing AHP techniques by interviewing experts to obtain the weight values and priority each criterion and sub criteria using pairwise comparison matrixes; (3) Applying GIS techniques using spatial data in the form of layers; (4) Identifying and prioritizing potential areas for TOD.

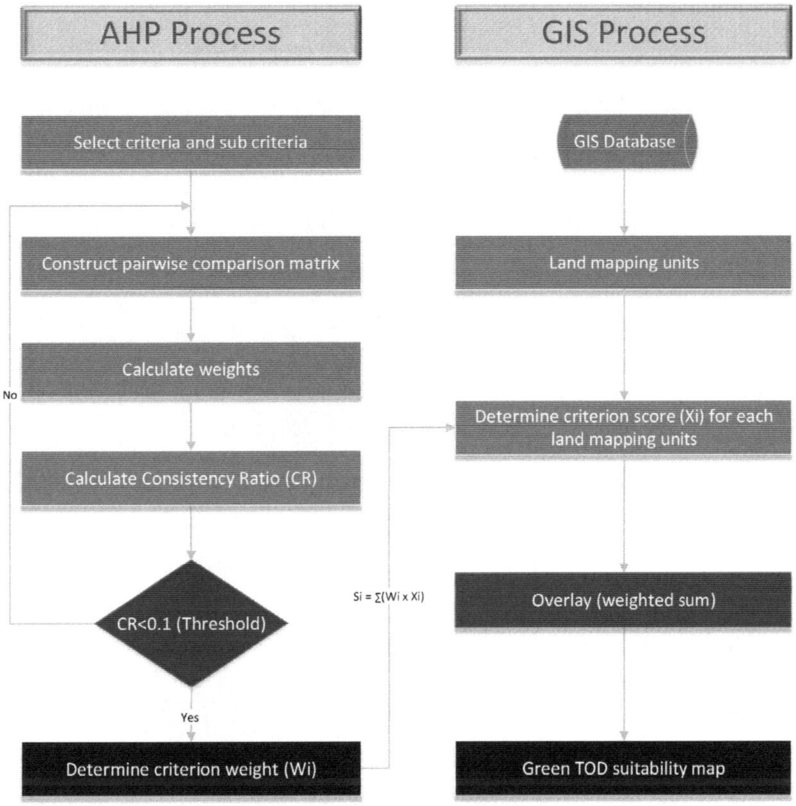

Figure 3. Methodology overview.

2.4. Analytical Hierarchy Process (AHP)

AHP is a method for making the rank and selection of the best decision alternative with several criteria. The most important thing in the AHP is the functional hierarchy with the main input of human perception. Functional hierarchy within groups can solve complex and unstructured problems. The calculation of the AHP method using pairwise comparison matrix is developed by Saaty [11] using the 1-9 scale. This scale is a useful matrix to obtain the quantitative value of a relative assessment.

Table 2. Saaty weighting scale.

Definition	Intensity of importance
Equally preferred	1
Moderately preferred	3
Strongly preferred	5
Very strongly preferred	7
Extremely preferred	9
Intermediate value between adjacent judgments	2,4,6,8

There are three main principles in problem-solving of AHP according to Saaty [11], namely: Decomposition, Comparative Judgment, and Logical Consistency. Understanding the AHP procedure includes the following stages: 1) Decomposition of the problem; 2) Assessment/weighting to compare elements; 3) Preparation of matrix and consistency test; 4) Prioritization on each hierarchy; 5) Synthesis of priorities; and 6) Decision-making.

Mathematical equations of AHP according to the formula below (Equation 1).

$$Cr = \frac{Ci}{ri} \quad (1)$$

Where *Cr* is consistency ratio index, *Ci* is consistency index, *ri* is random consistency index. While the consistency index (*Ci*) is derived using the following formula (Equation 2).

$$Ci = \frac{\lambda\ max - n}{n - 1} \quad (2)$$

Where *n* is the criteria number and $\lambda\ max$ is the maximum value of eigenvector.

2.5. Z-score method for standard score of criteria

It is used to know more detail where the position of a score in a distribution. The position in a distribution itself is denoted by the +/- symbol indicating that if the positive is above the mean and the negative denotes otherwise. The equation of Z-score is given below (Equation 3).

$$Z = (x-\mu) / \sigma \quad (3)$$

Where; μ = mean, x = score, σ = standard deviation

2.6. Land suitability evaluation

Land suitability evaluation for green TOD is the process of assessing the potential class for determining the potential area of TOD. The classification of land suitability is done by comparing the requested requirements by the characteristic of applied land. This method was adopted from FAO (Table 4). The results of it can be used as the basis for optimal and sustainable land.

Table 3. Land suitability classes of FAO.

Class	Name	Definition
S1	Highly Suitable	Land having no significant limitations to sustained application of a given use, or only minor limitations that will not significantly reduce productivity or benefits and will not raise inputs above an acceptable level.
S2	Moderately Suitable	Land having limitations which in aggregate are moderately severe for sustained application of a given use; the limitations will reduce productivity or benefits and increase required inputs to the extent that the overall advantage to be gained from the use, although still attractive, will be appreciably inferior to that expected on Class S1 land.
S3	Marginally Suitable	Land having limitations which in aggregate are severe for sustained application of a given use and will so reduce productivity or benefits, or increase required inputs, that this expenditure will be only marginally justified
N1	Currently Not Suitable	Land having limitations which may be surmountable in time but which cannot be corrected with existing knowledge at currently acceptable cost; the limitations are so severe as to preclude successful sustained use of the land in the given manner.
N2	Permanently Not Suitable	Land having limitations which appear so severe as to preclude any possibilities of successful sustained use of the land in the given manner.

The land suitability assessment was obtained by multiplying the suitability scores each factor involved (i.e, each raster cell on the map) with the factor weight. Calculation of suitability index was stored in one column. Integrating both the spatial components and the suitability index produced a land suitability map. The equation 4 is a formula of the land suitability assessment.

$$Si = \sum_{i=1}^{n} (\,Wi\;X\;Ri) \qquad (4)$$

Where;
Wi = the result of multiplication of all related weights in the factor hierarchy level
Ri = the standard assessment of each pixel assigned to the class on the criteria map
n = the number of criteria under element.

3. Result

3.1. Evaluation criteria

Evaluation criteria cover urban environment, economic performance, facilities services, transportation-related. 20 sub-criteria were selected according to the expert's experience, opinion, and information from various sources. The acquisition of knowledge was achieved through discussions with related field experts, authentic literary surveys, and analysis of historical data.

Table 4. Criteria and sub-criteria of evaluation.

Main criteria	Sub-criteria	Unit	Suitable rating				
			Suitable			Not suitable	
			High S1	Moderate S2	Marginal S3	Current N1	Permanent N2
Urban environment	BCR (building coverage ratio)	percentage	>80%	61-80%	41-60%	20-40%	<20%
	Park	percentage	2.55-9.08%	1.26-2.54%	0.48-1.25%	0.12-0.47%	0.00-0.11%
	Planned housing	ratio	1.23-2.32	0.72-1.22	0.32-0.71	0.08-0.31	0.00-0.07
	Residential	ratio	1.98-4.62	0.92-1.97	0.44-0.91	0.14-0.43	0.00-0.13
	FAR (floor area ratio)	FAR	4	3	2	1	0
Economic development	Population	density	28607-49345	14808-28606	6967-14807	2003-6966	0-2002
	CBD	density	0.000001	0.000001-0.000000	0.000001-0.000000	0.000001-0.000000	0
	Industrial	ratio	4.77-8.78	2.32-4.76	1.05-2.31	0.31-1.04	0.00-0.30
	Commercial and business	ratio	0.33-0.54	0.03-0.08	0.09-0.18	0.19-0.32	0.33-0.54
	Real estate	ratio	6.90-12.91	3.24-6.89	1.62-3.23	0.48-1.61	0.00-0.47
Facilities services	Education and public facility	ratio	0.37-0.81	0.009-0.36	0.004-0.008	0.01-0.003	0
	Recreation facility	ratio	0.50-0.86	0.23-0.49	0.05-0.22	0.01-0.04	0
	Hospital	distance	<250 m	250-500 m	500-800 m	800-1000m	>1000 m
	School	distance	<250 m	250-500 m	500-800 m	800-1000m	>1000 m
Transportation-related	Airport and Seaport	distance	<250 m	250-500 m	500-800 m	800-1000m	>1000 m
	Terminal	distance	<250 m	250-500 m	500-800 m	800-1000m	>1000 m
	Busway shelter	distance	<250 m	250-500 m	500-800 m	800-1000m	>1000 m
	Main road	distance	<250 m	250-500 m	500-800 m	800-1000m	>1000 m
	Tollgate	distance	<250 m	250-500 m	500-800 m	800-1000m	>1000 m
	Transportation facility	distance	<250 m	250-500 m	500-800 m	800-1000m	>1000 m

3.2. Assessment of pairwise comparisons and calculation of priorities

The AHP method was chosen for this study with the aim of being able to determine the relative importance of all selection criteria. The total suitability value gained through this process is the result of a linear combination of suitability scores from each criterion on each unit of land that is displayed in raster cell form on the map layer. The AHP method is generally implemented using the pair wise comparison technique that is a mathematical calculation that simplifies the rank preference among the proposed criteria. The computation matrix of pairwise comparison along with the calculation phases is described in the Table 5.

Table 5. Pairwise comparisons and assessment of criteria.

Criteria	Experts												SUM	Weighting	Ranking
	E1	E2	E3	E4	E5	E6	E7	E8	E9	E10	E11	E12			
Transportation-related	0.53	0.45	0.50	0.41	0.41	0.52	0.52	0.52	0.26	0.26	0.26	0.12	4.75	**0.40**	1
Economic development	0.08	0.08	0.07	0.38	0.38	0.26	0.26	0.26	0.48	0.48	0.48	0.49	3.68	**0.31**	2
Facilities services	0.25	0.27	0.26	0.12	0.13	0.14	0.14	0.14	0.14	0.14	0.14	0.18	2.05	**0.17**	3
Urban environment	0.14	0.20	0.17	0.09	0.08	0.08	0.08	0.08	0.13	0.13	0.13	0.20	1.51	**0.13**	4
inconsistency	*0.08*	*0.06*	*0.06*	*0.09*	*0.05*	*0.00*	*0.00*	*0.00*	*0.00*	*0.00*	*0.00*	*0.08*			

No	Main-criteria	Weight	Sub-criteria	Experts												SUM	Weight	Rank-ing
				E1	E2	E3	E4	E5	E6	E7	E8	E9	E10	E11	E12			
I	**Transportation-related**	**0.40**	Terminal	0.22	0.21	0.29	0.22	0.19	0.41	0.41	0.41	0.13	0.13	0.13	0.40	3.16	0.26	1
			Transportation facility	0.09	0.08	0.14	0.33	0.32	0.16	0.16	0.16	0.41	0.41	0.41	0.25	2.89	0.24	2
			Main road	0.12	0.13	0.14	0.09	0.10	0.21	0.21	0.21	0.11	0.11	0.11	0.10	1.65	0.14	3
			Airport and seaport	0.28	0.33	0.09	0.08	0.10	0.10	0.10	0.10	0.15	0.15	0.15	0.12	1.74	0.14	4
			Busway shelter	0.18	0.15	0.24	0.17	0.16	0.07	0.07	0.07	0.13	0.13	0.13	0.07	1.58	0.13	5
			Tollgate	0.11	0.11	0.11	0.12	0.14	0.05	0.05	0.05	0.07	0.07	0.07	0.06	0.98	0.08	6
			inconsistency	*0.09*	*0.09*	*0.08*	*0.08*	*0.09*	*0.03*	*0.03*	*0.03*	*0.06*	*0.06*	*0.06*	*0.09*			
II	**Economic develop-ment**	**0.31**	Population density	0.45	0.41	0.34	0.40	0.29	0.46	0.46	0.46	0.44	0.44	0.44	0.23	4.82	0.40	1
			CBD	0.35	0.24	0.25	0.27	0.18	0.25	0.25	0.25	0.21	0.21	0.21	0.19	2.85	0.23	2
			Industrial	0.14	0.11	0.14	0.20	0.33	0.15	0.15	0.15	0.18	0.18	0.18	0.23	2.15	0.17	3
			Commercial and business	0.23	0.19	0.18	0.05	0.08	0.07	0.07	0.07	0.06	0.06	0.06	0.13	1.25	0.10	4
			Real estate	0.08	0.06	0.09	0.07	0.11	0.07	0.07	0.07	0.11	0.11	0.11	0.23	1.18	0.10	5
			inconsistency	*0.09*	*0.08*	*0.06*	*0.05*	*0.08*	*0.04*	*0.04*	*0.04*	*0.01*	*0.01*	*0.01*	*0.09*			
III	**Facilities services**	**0.17**	School	0.08	0.05	0.08	0.09	0.23	0.59	0.59	0.59	0.50	0.50	0.50	0.24	4.06	0.34	1
			Education and public facility	0.48	0.53	0.42	0.41	0.52	0.20	0.20	0.20	0.23	0.23	0.23	0.14	3.78	0.32	2
			Recreation facility	0.25	0.28	0.33	0.38	0.16	0.08	0.08	0.08	0.08	0.08	0.08	0.36	2.26	0.19	3
			Hospital	0.20	0.14	0.16	0.12	0.09	0.13	0.13	0.13	0.19	0.19	0.19	0.26	1.90	0.16	4
			inconsistency	*0.05*	*0.06*	*0.08*	*0.09*	*0.06*	*0.02*	*0.02*	*0.02*	*0.01*	*0.01*	*0.01*	*0.08*			
IV	**Urban environ-ment**	**0.13**	FAR	0.07	0.11	0.09	0.49	0.49	0.48	0.48	0.48	0.28	0.28	0.28	0.48	4.03	0.34	1
			BCR	0.36	0.33	0.21	0.31	0.31	0.10	0.10	0.10	0.11	0.11	0.11	0.13	2.28	0.19	2
			Formal housing	0.18	0.18	0.22	0.06	0.06	0.17	0.17	0.17	0.21	0.21	0.21	0.16	2.00	0.17	3
			Informal housing	0.13	0.14	0.17	0.09	0.09	0.12	0.12	0.12	0.26	0.26	0.26	0.13	1.90	0.16	4
			Park	0.26	0.24	0.31	0.05	0.05	0.13	0.13	0.13	0.13	0.13	0.13	0.11	1.78	0.15	5
			inconsistency	*0.09*	*0.09*	*0.06*	*0.09*	*0.09*	*0.07*	*0.07*	*0.07*	*0.05*	*0.05*	*0.05*	*0.01*			

3.3. Land suitability map of green TOD

The spatial pattern of potential suitability areas of TOD is shown in Figure 4. The spatial pattern formed showed that the moderate suitability area (yellow colour) was randomly distributed almost all regions, the high suitability area (red colour) indicated clustering pattern in the middle of study area and the pattern of low suitability area (blue colour) showed clustered pattern by dominating the entire study area.

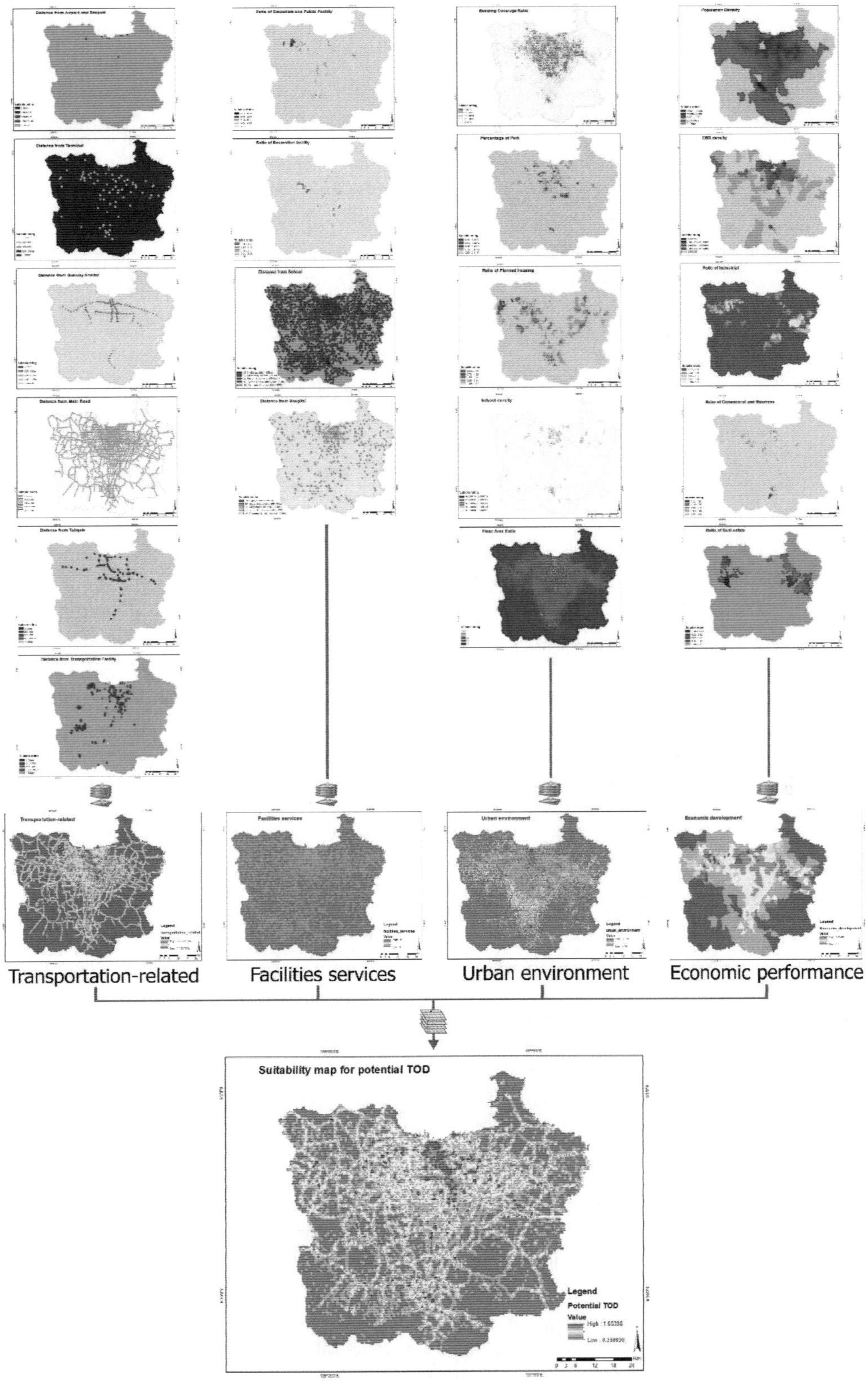

Figure 4. Suitability map of green TOD.

3.4. The distribution land of potential green TOD

Comparison of land suitability analysis (Table 6 and Figure 5) shows that the highest class is permanently not suitable (36,4%) with the area of 2769.68325 ha. While the lowest class is highly suitable (2.3%) with the area of 174.69302 ha. Moderately suitable (9%) with the area of 682.65415 ha can be used as an alternative area to be a developing area of potential green TOD.

Table 6. The distribution land of potential green TOD.

Suitable rating	class	percentage (%)	area (ha)
Highly suitable	S1	2.3	*174.69302*
Moderately suitable	S2	9.0	*682.65415*
Marginally suitable	S3	28.1	*2135.25561*
Currently not suitable	N1	24.3	*1847.69085*
Permanently not suitable	N2	36.4	*2769.68325*
Total		*100*	*7609.97688*

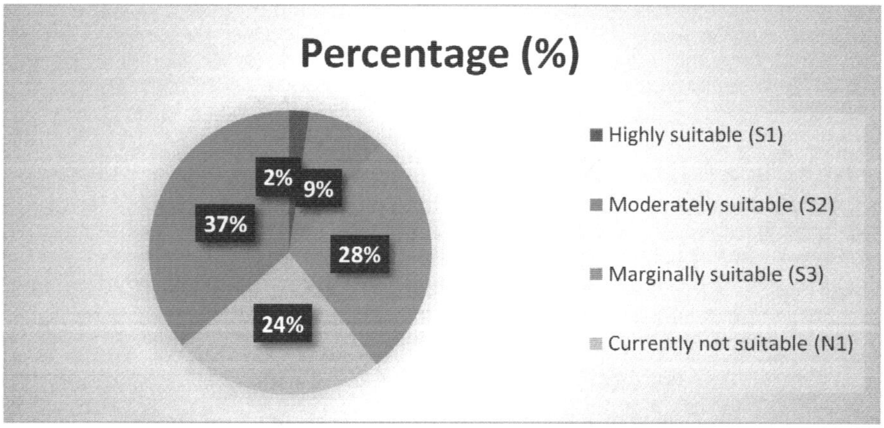

Figure 5. Chart of distribution land.

4. Discussion

The mobility of inhabitants in the JMR is a daily commuting to job places in Jakarta City, hence transportation modes and transit areas should be integrated, not only limited in administrative boundaries but also it could cope congestion, pollution, cost, and accidents. Development and improvement of transportation services could be achieved by implementing effective and efficient organizational governance, especially the management of transit areas such as TOD.

Moreover, JMR inhabitants are right now faced with the challenge of getting occupied under the circumstances of green open area. JMR especially the Jakarta City currently has low availability of green open spaces. Seeing the longing of the inhabitants would be dwelling which surrounded by a beautiful environment. TOD presents the green concept in which indicates not only by less emission of vehicles, since the community uses the last-mile travel by walking, from their homes; but also, TOD provides a strategic transit location with green area, consisting of gardens, pedestrian and cycling track, as well as easy access to the office area with a variety of commercial areas [12].

The potential area of green TOD development in JMR (highly and moderately suitable) based on this study has an area of 857 hectares or 11.3% of the total JMR area. It is concentrated in the central region that is Jakarta capital and surrounding areas. This TOD, if developed with the green concept, will reflect the typical characteristics of Indonesian heritage [13], and offers a green environment that is organized together with the natural elements around it. TOD of the railway station is located close to the outer ring road and toll gate, equipped with the planned Light Rail Transit (LRT) as well as direct access to international airport

The rapid growth of office and industrial areas as well as the increasing infrastructure in the suburbs (Botabek) to Jakarta City and vice versa, significantly makes Botabek a buffer area for favorite choice as the housing location. Likewise, the changes in the Spatial law/RTRW due to special circumstance should be paid consideration by the municipality. Accommodating those needs, the government policy is trying to develop the TOD project, which will be a strategic area and facilitate the needs of green residential for the employee to work in the office or industrial area and its surrounding.

The next study for future research, there is an endeavour to 'retrofit' with a decision-support system (DSS) modelling with a 'what-if' capability, which relates to the GIS-based MCE Analytics above, and the carrying capacity raster calculation feature in GIS layers (economic exploitation vs environmental protection) of the surrounding locations of the TODs. Our recommendation proposed TOD as the foundation to work with the Smart-City Movement and Urbanism, where smart-mobility will use Internet-of-Things (IoT), transportation sensors, automations, robotics, early warning systems for safety, security and disaster management. This will almost make sure to erase fossil-based fuel vehicles for the advent of full electric vehicle (EV) environment.

5. Conclusion

The potential area of green TOD is identified based on the calculation of selected criteria to produce the desired condition. The spatial planning of TOD for JMR makes this transitional area is a service center with mixed functions and modal transfer shelters. JMR is expected to be a model of sustainable development through the multifunctionality services of social and regional facilities, therefore rapid commuting can be optimally served within an integrated service area.

References

[1] Taki H M 2017 *J. Geomatics Plan.* **4** 12726
[2] Hernbäck J 2012 *Influence of Urban Form on Co-presence in Public Space: A Space Syntax Analysis of Informal Settlements in Pune, India* Master Thesis (Stockholm, Sweden: KTH, Royal Institute of Technology)
[3] Karimi K, Parham E and Acharya A 2015 *Proceedings of the 10th Int. Space Syntax Symp. (SSS10)*, London, UK
[4] Taki H and Lubis M 2017 *J. Appl. Geospatial Inf.* **1** 36-43
[5] Taki H M, Maatouk M M H and Qurnfulah E M 2017 *J. Appl. Geospatial Inf.* **1** 26–35
[6] Bunruamkaew K and Murayam Y 2011 *Procedia Soc. Behav. Sci.* **21** 269–78
[7] Malczewski J 2006 *Int. J. Appl. Earth Obs.* **8** 270-7
[8] Singh Y, Fard P, Zuidgeest M, Brussel M and van Maarseveen M 2014 *J. Transp. Geogr.* **35** 130-43
[9] Fard P 2013 *Measuring Transit Oriented Development: Implementing a GIS-Based Analytical Tool for Measuring Existing TOD Levels* Master Thesis (Enschede, the Netherlands: University of Twente)
[10] Taki H M, Maatouk M M H, Qurnfulah E M and Aljoufie M O 2017 *J. Geosci. Eng. Environ. Technol.* **2** 84-94
[11] Saaty T L 1990 *Eur. J. Oper. Res.* **48** 9-26
[12] Cervero R and Sullivan C 2011 *Int. J. Sustain. Dev. World Ecol.* **18** 210–8
[13] Shao G, Li F and Tang L 2011 *Int. J. Sustain. Dev. World Ecol.* **18** 187-9

The characteristic of Natural Organic Matter (NOM) of water from Cikapundung River Pond

D Roosmini, S Notodarmojo and M R Sururi*

Environmental Engineering Department, Institut Teknologi Bandung, JL. Ganesha No.10, Bandung, Indonesia

*Email: rangsoer@gmail.com

Abstract. Natural organic matter (NOM) is an important parameter in drinking water due to its roles as one of main precursors of trihalomethanes (THMs). The characteristic of NOM might be influenced by land use of catchment area. The present study aims to analyse the correlation between various NOM parameters and THMs, and physical characteristics of raw water. The effect of land use around the catchment is also presented. The samples were taken from Cikapundung river pond, which is used by the water company of Bandung City as raw water sources for drinking water. Geographic Information System (GIS) was applied to distinguish anthropogenic and natural land use around the water catchment. The samples were categorised as unfiltered samples and filtered samples to measure dissolved material. Turbidity and conductivity were measured to obtain the physical characteristics of water. NOM parameters were as of TOC, DOC, COD, UV_{254} and UV_{355}. The turbidity and conductivity of water were 5-34.7 NTU and 132-306 μmhos/cm, respectively. COD was from 32 to 76.8 mg/L. TOC was between 4.22 and 6.92 mg/L. UV_{254} and UV_{355} ranged 0.15-0.38 cm^{-1} and 0.09-0.16 cm^{-1}, respectively. COD and conductivity correlated well in unfiltered ($r = 0.923$) and filtered sample ($r = 0.995$). In both unfiltered and filtered samples, UV_{254} were correlated with UV_{355} with r of 0.986 and 0.995 respectively. The results revealed that catchments area of the pond was possibly dominated by anthropogenic activities.

1. Introduction

Natural organic matter (NOM) is an important water quality parameter. NOM has been known as the main precursor of disinfection by products (DBPs) in chlorination process. USEPA [1] identified that the NOM is the major precursor of two out of three DBPs such as organic oxidation by product and halogenated organic by product. Jacangelo *et al* [2] found that NOM in the pretreated water may generate serious problems to treatment process of drinking water.

In chlorination process, the most popular DBPs is trihalomethanes (THMs). THMs are potentially considered as carcinogenic, mutagenic, and able to damage liver function [3]. The four THM species, including trichloromethane (TCM), and the brominated compounds such as bromodichloromethane (BDCM), dibromochloromethane (DBCM), and tribromomethane (TBM) are mostly found in disinfected water [3].

The heterogeneity of NOM is influenced by the location of water sources and land use of the catchment area [4-9]. Due to the heterogeneity of NOM, not all parameters of NOM are correlated with THMs. Previous studies suggested that chromophoric NOM parameters such as UV_{254} is correlated with THMs [10-15]. Other studies concluded that total organic carbon (TOC) and dissolved organic carbon

(DOC) correlated with THMs [16-18]. However, in Indonesia, chemical oxygen demand (COD) is the most common parameter to determine organic matter in raw or pretreated water quality. The problem arises since there is no strong correlation between COD and THMs.

Turbidity and conductivity are common physical parameters and relatively easy to be measured. Further, these parameters can be used as surrogate parameters, thus it is important to know the correlation between these parameters and NOM parameters. In Indonesia, the studies on correlation among NOM parameters and physical parameters is rare.

Cikapundung River has been the primary raw water sources of PDAM Tirtawening, a local water company in Bandung City. Although this river has been determined as a strategic area of the city, the condition of Cikapundung River has been severely polluted. The Environmental Protection Agency (EPA) of Bandung City accounted the water quality status by STORET method in the Cikapundung river and they found the river is polluted. Another important role of Cikapundung River is to provide raw water as many as 640 L/sec in water production region 2 of PDAM Tirtawening. One of the installations is the Mini Plant Dago Bengkok, which is used by the river pond to provide detention time for discrete particle to settle.

It is important to study the characteristic of NOM from river pond because they are so many installations in Indonesia, which use river pond as the main water source. It is believed that the pond can provide raw water better in quality but there is no sufficient information about the characteristic of NOM and physical characteristic of water at water pond. Furthermore, Water Company has been used chlorine in disinfection process. Chlorine is the most used chemical oxidant for drinking water disinfection in Indonesia.

This research is aimed to analyze the characteristic of physical parameters such as turbidity and conductivity, lability NOM such as COD, parameters of NOM which related to THMs (UV_{254}, TOC and DOC), and terrestrial NOM (UV_{355}). This study evaluated the correlation among those parameters on unfiltered and filtered samples from the selected river pond.

2. Material and Methods

2.1. *Study area*

The Cikapundung is important river since it has many functions, such as drinking water source, key drainage canal system, and recreation. The Cikapundung River has provided raw water for three main installations of PDAM Tirtawening Kota Bandung. The sampling location coordinate of Cikapundung Pond is S 06°51'33.1" & E 107°38.57'57.1". The pond provides 40 L/sec of raw water to be treated at mini plant and the treated water is distributed to North area of Bandung.

2.2. *Geographic Information System (GIS) work*

GIS were used to delineate the catchment area of the river pond and determine the natural area and anthropogenic areas which were affected by human activities in the catchment area of river pond. The shape file map was obtained from the Department of Spatial Planning and Human Settlement of West Java Province.

2.3. *Water sample collection*

The water sampling was conducted in 10 days at the outlet of river pond. The samples were divided into unfiltered and filtered sample. To determine dissolved NOM, the samples were filtered by Advantech membrane with diameter of 0.45 μm.

2.3.1. Physical characteristic measurements. The measurement method of these parameters were based on Standard Method and listed at Table 1.

Turbidity is an appearance of the optical properties that causes light to be scattered and absorbed rather than transmitted [19]. The measurement method of turbidity is by nephelometric method

IOP Conf. Series: Earth and Environmental Science **160** (2018) 012021 doi:10.1088/1755-1315/160/1/012021

according standard method 2130 B. Conductivity is a measure of the ability of an aqueous solution to carry an electric current [19]. The measurements of these parameters were carried out *in situ*.

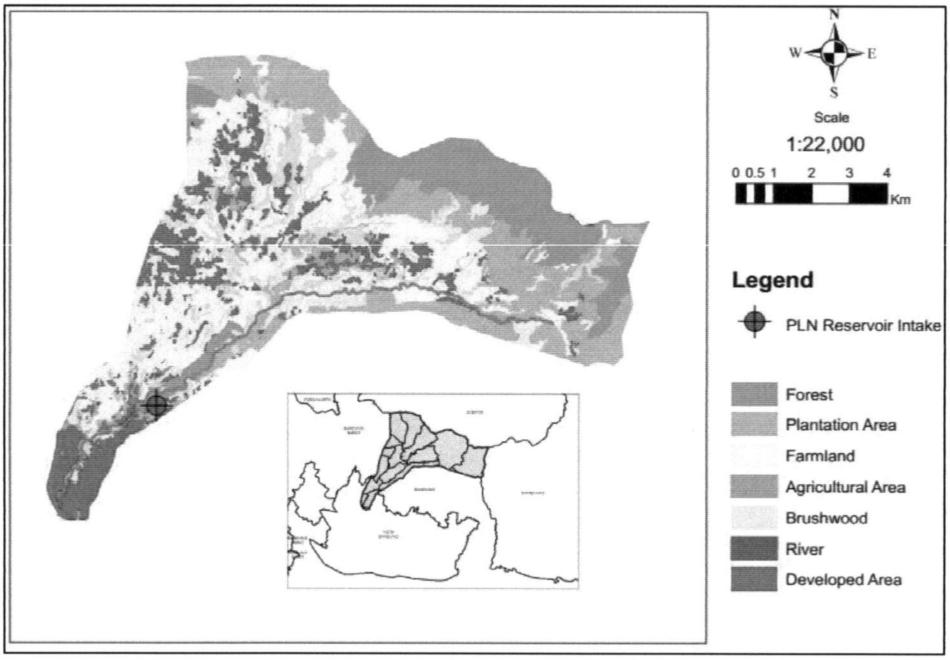

Figure 1. Catchment area of Upper Cikapundung.

Table 1. Method of measurements.

No	Parameter	Measurement method
1	Turbidity	Standard methods 2130 B
2	Electrical Conductivity	Standard methods 2510
3	COD	Standard methods 5220 D
4	DOC	Standard methods 3510 B
5	UV_{254}	Standard methods 5910 B

2.3.2. NOM measurements. UV absorption is a useful surrogate parameters of organic contaminant in fresh water [19]. UV_{254} measures the humic, lignin, tannin and various aromatic compound [19]. The origins of NOM influence THMs formation [15]. Some parameters such as UV_{355} is described as the chromophoric NOM derived from terrestrial [20]. COD is a measurement of the oxygen required to oxidize chemically soluble and particulate organic matter in water [21].

Chromophoric NOM parameters such as UV_{254} and UV_{355} were measured by Spectrophotometer UV-VIS CAMSPEC M500. TOC and DOC were measured by TOC analyzer SIEVER Innovox 0545 and COD measurement used close reflux method.

2.4. Statistical analysis

Correlation is a statistical technique that indicates how strong pairs of NOM and physical parameters are related. The NOM parameters were: lability NOM (COD), chromophoric NOM (UV_{254} and UV_{355}), and bulk organic carbon NOM (TOC and DOC). Meanwhile the physical parameters were turbidity and electrical conductivity. The data were analyzed statistically by SPSS 19.

IOP Conf. Series: Earth and Environmental Science **160** (2018) 012021 doi:10.1088/1755-1315/160/1/012021

3. Result and discussion

3.1. Land use pattern in catchment area of river pond

The river pond is part of Cikapundung River. Figure 2 illustrates the catchment area of river pond. Land use in the catchment area was grouped into two categories:

1. Forest is natural area (green), there were no significant human activities in this area,
2. Developed area (yellow), there were significant human activities in these areas. The land use in this area are: agricultural; farm; livestock; traditional market; and residential area.

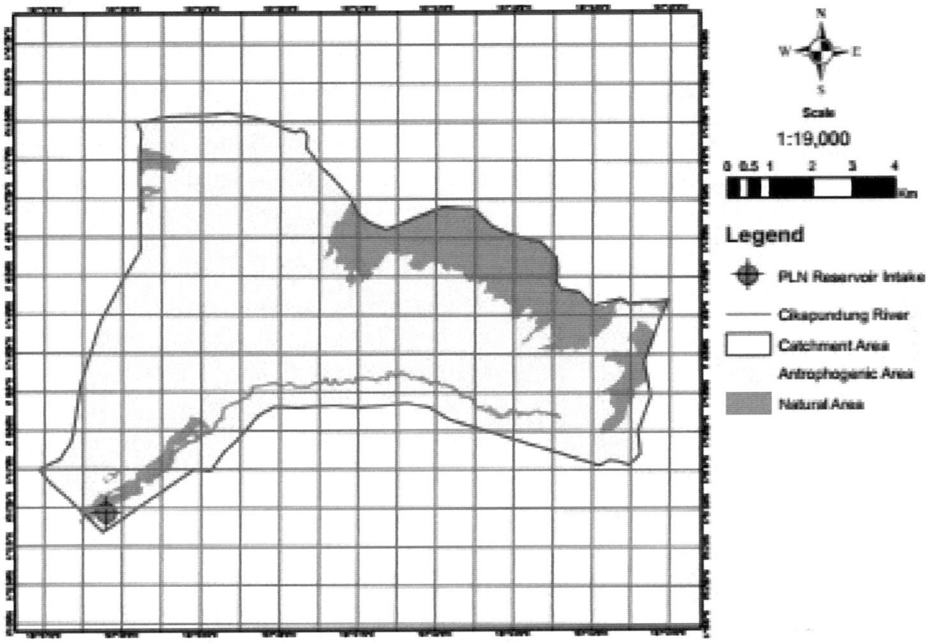

Figure 2. Land use pattern in catchment area of the intake.

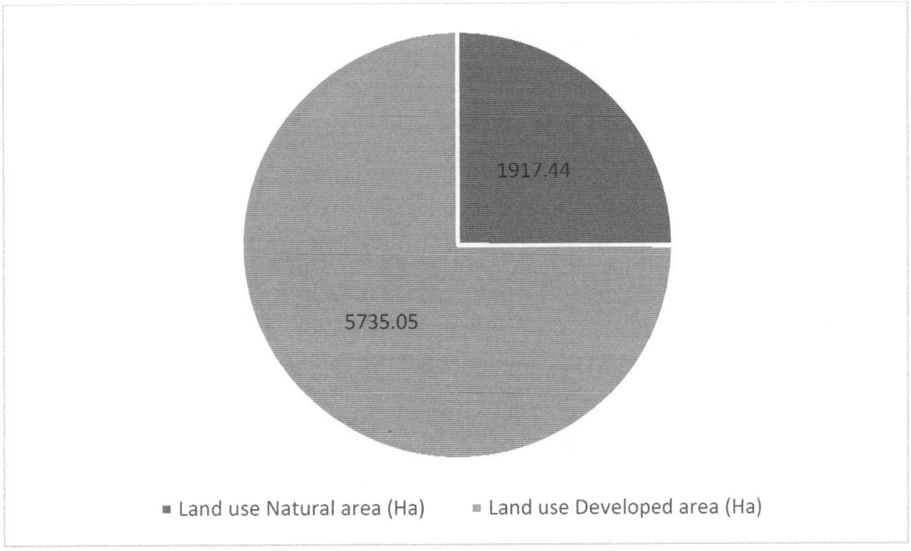

Figure 3. Natural and anthropogenic area in the catchment area of river pond.

Figure 3 shows, the catchment area of river pond was predominantly used for anthropogenic activities (75%). Meanwhile the natural area was only 25%. This condition affected the water quality of Cikapundung River. The human activities which affected the water quality in this catchment area are (PDAM Tirtawening, 2015):

- Some farming where livestock activities at the catchment area directly discharge the waste to the rivers,
- Domestic and non-domestic activities that discharge wastewater and solid waste directly to the river.

3.2. Physical and NOM characteristic

Table 2 summarizes the physical characteristic of filtered and unfiltered samples. In the unfiltered samples, the turbidity of samples ranged between 15.8 and 34.7 NTU. However, the turbidity decreased to 0.1-4.3 NTU in filtered samples, and it was suitable with standard of drinking water quality standard (<5 NTU). This condition happened because the turbidity of the water sample was dominated by suspended and colloid material, thus the turbidity of the sample decrease significantly after the sample was filtered. Furthermore, the sampling was done during rainy seasons which may increase the concentrations of suspended particles in the sample.

Table 2. Physical characteristic of the sample water from Cikapundung River Pond.

Sample	Turbidity (NTU)	Conductivity (μmhos/cm)
Unfiltered	15.8-34.7	132-306
Filtered	0.1-4.3	82-201

Electrical conductivity of water is an indicator of its salinity or total salt content [22]. The conductivity of sample ranged between 132 and 306 μmhos/cm. After the sample was filtered, it decreased by 37%, ranged between 82 and 201 μmhos/cm. The conductivity of the sample is predominantly caused by dissolved material. This was due to dissolved solid in the water conduct electrical current in the water [23]. Normally, in the surface water the conductivity decreased due to dilution with runoff water [24].

Table 3. Organic characteristic of the sample water.

Sample	COD (mg/L)	UV_{254} (cm^{-1})	UV_{355} (cm^{-1})	TOC (mg/L)
Filtered	32-76.8	0.15-0.38	0.09-0.21	4.22-6.92
Unfiltered	17-48	0.12-0.29	0.07-0.2	3.28-3.58

In Table 3, all COD values in unfiltered and filtered samples were above the maximum value of COD in stream standard (10 mg/L). These results show that more than half of lability NOM, measured as COD were dissolved. This was supported by Tommassen [23] which found COD comprise of particle suspended and dissolved solid in the water. This may indicate water pollution by organic compounds. GIS shows, the land use area of the catchment is dominated by anthropogenic activities. Direct discharge from anthropogenic activities affected this condition.

TOC were between 4.22 and 6.92 mg/L and the DOC were between 3.28 and 3.58 mg/L. The concentration of particulate organic carbon (POC) in the sample were between 0.94-3.34 mg/L, these value was lower than DOC. TOC consist of DOC and POC [5,25]. Niu *et al* [26] reported, DOC in Yuqiao Reservoir were between 2.96-3.54 mg/L and classified as medium. Meanwhile Zhao *et al* [14] reported the DOC from polluted water was only 2.02 mg/L, slightly lower from river pond sample.

Organic aromatic of the sample as UV_{254} was 0.15-0.38 cm^{-1} before filtered and decrease 23% after filtering. This organic value is considerably high. Hua *et al* [27] reported that in raw water polluted by agricultural activities, the UV_{254} was 0,313 cm^{-1}. Meanwhile, Wang *et al* [28] reported that the value of UV_{254} from filtered water of Songhua river was only 0.063 cm^{-1}. GIS shows that the developed area in the catchment of river pond were not only agricultural but also residential area, traditional market, livestock and farm. This fact shows the sample dominated by organic aromatic both in filtered and unfiltered water.

Organic from terrestrial as UV_{355} were recorded between 0.09-0.21 cm^{-1} in unfiltered and decreased by 22% in filtered water. The concentration level of UV_{355} was lower than organic aromatic (UV_{254}), which may show that allocthonous NOM from terrestrial is part of autochonous NOM in the water body. Furthermore, based on GIS, the catchment area of Cikapundung river pond is dominated by domestic activities and non-domestic activities. This may explain the high concentrations of NOM parameters and physical parameter of NOM in both filtered and unfiltered samples except for turbidity.

3.3. Correlation among physical characteristics and NOM in River Pond

Correlation among physical characteristics and NOM in Cikapundung River in unfiltered and filtered samples are shown in Table 4 and Table 5 respectively. There were strong correlations between COD and electrical conductivity in both filtered and unfiltered samples. Barakat *et al* [29] found similar result in Oum Er Rbia River which were polluted by agricultural and domestic activities. Conductivity and COD can be used as indirect chemical indicators of fecal contamination [30].

The correlation between COD and turbidity was strong but slightly lower than 95%. However, after filtering the correlation was insignificantly weak due to particle removal. The correlation between two parameters of chromophore NOM (UV_{254} and UV_{355}) in unfiltered sample and filtered sample were significant and strong. These were because the autochonous NOM in the river water was dominated by terrestrial organic matter. In addition, the location of river pond is in the national forest park of Djuanda thus organic matter from this forest discharge directly to this pond.

Table 4. Correlation among physical characteristics and NOM on unfiltered sample.

		Turbidity	Conductivity	COD	UV_{254}	UV_{355}	TOC
Turbidity	Pearson Correlation	1					
	Sig. (2-tailed)						
Conductivity	Pearson Correlation	.877	1				
	Sig. (2-tailed)	.051					
COD	Pearson Correlation	.753	.923*	1			
	Sig. (2-tailed)	.142	.025				
UV_{254}	Pearson Correlation	.804	.506	.314	1		
	Sig. (2-tailed)	.101	.384	.607			
UV_{355}	Pearson Correlation	.873	.583	.379	.986**	1	
	Sig. (2-tailed)	.054	.302	.530	.002		
TOC	Pearson Correlation	-.038	-.419	-.575	.152	.205	1
	Sig. (2-tailed)	.952	.483	.311	.807	.741	

* Correlation is significant at the 0.05 level (2-tailed). ** Correlation is significant at the 0.01 level (2-tailed).

Table 5. Correlation among physical characteristics and NOM on filtered sample.

		Turbidity	Conductivity	COD	UV_{254}	UV_{355}	DOC
Turbidity	Pearson Correlation	1					
	Sig. (2-tailed)						
Conductivity	Pearson Correlation	.004	1				
	Sig. (2-tailed)	.995					
COD	Pearson Correlation	.055	.896*	1			
	Sig. (2-tailed)	.930	.040				
UV_{254}	Pearson Correlation	-.505	-.819	-.808	1		
	Sig. (2-tailed)	.386	.090	.098			
UV_{355}	Pearson Correlation	-.434	-.868	-.831	.995**	1	
	Sig. (2-tailed)	.465	.056	.081	.000		
DOC	Pearson Correlation	-.134	.784	.601	-.406	-.473	1
	Sig. (2-tailed)	.830	.116	.283	.498	.421	

* Correlation is significant at the 0.05 level (2-tailed). ** Correlation is significant at the 0.01 level (2-tailed).

In this study, there were no significant correlation between bulk organic NOM such TOC and DOC as well as organic parameters. This was due to many organic compounds which moderately contain organic carbon did not correlate with organic carbon which absorb significantly UV wavelengths, and chemical oxygen demand which exceed the standard.

4. Conclusion

The land use of the catchment area of Cikapundung river pond is dominated by anthropogenic activities which was furthermore affected water quality. Dissolved material plays crucial role on NOM parameters and electrical conductivity. Meanwhile the turbidity of the water mainly caused by non-dissolved material. There were significant and strong associations between COD and electrical conductivity as well as UV_{355} and UV_{254} in both filtered and unfiltered samples.

References

[1] USEPA 1999 *Alternative Disinfectants and Oxidants Guidance Manual*
[2] Jacangelo J G, De Marco J, Owen D M and Randtke SJ 1995 *J. AWWA* **87** 64-77
[3] WHO 2004 *Guidelines for Drinking-Water Quality* (Geneva: World Health Organization)
[4] Scott M, Jones M, Woof C, Simon B and Tipping E 2001 *Env. Int.* **27** 449-62
[5] Thurman E M 1985 *Organic Geochemistry of Natural Waters* (Dordrecht: Springer Science & Business Media)
[6] Moyer R P, Powell C E, Gordon D J, Long J S and Bliss C M 2015 *Appl. Geochem.* **63** 550-62
[7] Sawicka K, Monteith D, Vanguelova E, Wade A and Clark J 2016 *Ecol. Indic.* **68** 36-51
[8] Yu S, Xu Z, Wu W and Zuo D 2016 *Ecol. Indic.* **60** 202-12
[9] Wilson H F, Saiers J E, Raymond P A and Sobczak W V 2013 *Ecosystems* **16** 604-16
[10] Reckhow D A, Singer P C and Malcolm R L 1990 *Env. Sci. Technol.* **24** 1655-64
[11] Ates N, Kaplan S S, Sahinkaya E, Kitis M, Dilek FB and Yetis U 2007 *J. Hazard. Mat.* **142** 526-34
[12] Pifer A D and Fairey J L 2014 *Env. Eng. Sci.* **31** 117-26

[13] Ma D, Peng B, Zhang Y, Gao B, Wang Y, Yue Q and Li Q 2014 *Bioresource Technol.* **165** 81-7
[14] Zhao Z-Y, Gu J-D, Fan X-J and Li H-B 2006 *J. Hazard. Mat.* **134** 60-6
[15] Awad J, Van Leeuwen J, Chow C, Drikas M, Smernik R J, Chittleborough D J and Bestland E 2016 *J. Hazard. Mat.* **308** 430-9
[16] Chowdhury S and Champagne P 2008 *Global Nest J.* **10** 80-91
[17] Nguyen H V-M, Lee M-H, Hur J and Schlautman M A 2013 *J. Hydrol.* **481** 132-42
[18] Tubic A, Agbaba J, Dalmacija B, Molnar J, Maletic S, Watson M and Perović S U 2013 *J. Env. Manag.* **118** 153-60
[19] APHA 2005 *Standard Methods for the Examination of Water and Wastewater* (Washington, DC: American Public Health Association)
[20] Dainard P G and Guéguen C 2013 *Mar. Chem.* **157** 216-23
[21] Sawyer C N and McCarty P L 1978 *Chemistry for Environmental Engineers* (New York: McGraw-Hill Book Company)
[22] Morrison G, Fatoki O, Persson L and Ekberg A 2001 *Water SA* **27** 475-80
[23] Tommassen G 2014 *On the Correlation Between Turbidity, Conductivity and COD* (Delft: TU Delft)
[24] Muller F L and Tankéré-Muller S P 2012 *Sci. Total Env.* **435** 351-62
[25] Pagano T, Bida M and Kenny J E 2014 *Water* **6** 2862-97
[26] Niu Z-G, Wei X-T and Zhang Y 2015 *Env. Sci. Pollut. Res.* **22** 17508-17
[27] Hua G, Reckhow D A and Abusallout I 2015 Chemosphere **130** 82-9
[28] Wang H, Liu D M, Zhao Z W, Cui F Y, Zhu Q and Liu T M 2010 *J. Zhejiang Univ. Sci. A* **11** 143-50
[29] Barakat A, El Baghdadi M, Rais J, Aghezzaf B and Slassi M 2016 *Int. Soil Water Conserv. Res.* **4** 284-92
[30] Howard I, Espigares E, Lardelli P, Martin J and Espigares M 2004 *Env. Toxicol.* **19** 241-9

Total arsenic in tuna fish candidate reference material preparations: Homogeneity and stability testing

T A Koesmawati[1*], Buchari[2] and J Ng[3]

[1] Research Unit for Clean Technology, Indonesian Institute of Sciences (LIPI), Komplek LIPI Bandung, Jalan Sangkuriang, Gedung 50, Bandung, 40135, Indonesia
[2] Analytical Chemistry Department, Bandung Institut of Technology (ITB), Jalan Ganeca 10, Bandung 40135, Indonesia
[3] The University of Queensland, National Research Centre for Environmental Toxicology (Entox), Brisbane 4108, Queensland, Australia

[*]E-mail: tiny.agustini@gmail.com

Abstract. Reference material plays an important role in analytical measurements. Total arsenic was analyzed in tuna fish samples using three different preparation methods for stability. The stability of total arsenic in tuna fish was evaluated over 12 months, under room temperature storage condition. Each measurement was conducted in triplicate in three different laboratories. Statistical analysis was carried out using Student's t-test. Two different methods were conducted to confirm homogeneity of the candidate reference material in this study. Random bottle number was applied in this analysis. Statistical analysis was carried out and the homogeneity for total arsenic was calculated using the one-way analysis of variance (ANOVA). The statistical results showed no significant changes in stability and the sample was homogeneous. The candidate reference material developed in this study demonstrated its suitability for quality assurance of the total arsenic measurement in tuna fish.

1. Introduction

Arsenic is a metalloid widely distributed in the earth's crust and present at an average concentration of 2 µg/g. Arsenic can exist in four valences states: –3, 0, +3 and +5. Under reducing conditions, arsenite (As (III)) is the dominant form; arsenate (As (V)) is generally the stable form in oxygenated environments. Arsenic causes adverse health effects including cancers in human. At present, millions of people worldwide suffer from chronic arsenic poisoning [1,2] mainly due to consumption of arsenic-contaminated water and food. Long-term exposure to arsenic in drinking-water is causally related to increased risks of cancer in the skin, lungs, bladder and kidney, as well as other skin changes such as hyperkeratosis and pigmentation changes. Increased risks of lung and bladder cancer and of arsenic-associated skin lesions have been reported to be associated with ingestion of drinking-water at concentrations 50 µg arsenic/L. Occupational exposure to arsenic, primarily by inhalation, is causally associated with lung cancer. Exposure–response relationships and high risks have been observed. Increased risks have been observed at cumulative exposure levels ≥ 0.75 (mg/m^3) × year (e.g. 15 years of exposure to a workroom air concentration of 50 µg/m^3).

Various of instrumental techniques for arsenic determination is available such as atomic absorption spectrometry (AAS), atomic fluorescence spectrometry (AFS), inductively coupled plasma atomic emission spectrometry (ICP-AES), inductively coupled plasma mass spectrometry (ICP-MS) and voltammetry. ICP-MS as an example, can serve as element-specific detectors when coupled to chromatographic separation techniques such as HPLC and GC. The use of CRM on testing and calibration process is a requirement from ISO/IEC 17025-2005. This material is an essential tool to the quality control of measurement methods [3]. The certification of a reference material is carried out according to the requirements of ISO Guide 34 series [4]. The ISO Guide 35 states that the certification process of a CRM requires a careful study of all sources of uncertainty that can cause impact on the validity of the certified values. In general, these sources are relevant to characterization uncertainty, homogeneity uncertainty, uncertainty stability inherent to transport, uncertainty stability inherent to the storage, that are essential for the development and certification of a CRM [5]. The homogeneity study is necessary in the certification process of a lot of a RM to demonstrate that the units of this lot are sufficiently homogeneous amongst themselves. The homogeneity should be evaluated between different units of the candidate CRM lot (bottles at this case) and also within the same unit.

Total arsenic in tuna fish is usually found in relative very low concentrations of below 10 μg/g. Two major difficulties in the measurement of total arsenic in tuna fish are the concentration and its matrix interferences [6–9]. Accuracy and precision in its measurement is mandatory for an accredited testing laboratory, therefore the availability of a suitable reference material (RM) is necessary [10,11]. RM is a material or substances with one or more of its properties being sufficiently homogeneous and stable that are well established to be used for calibration of an apparatus, the assessment of a measurement method, or for assigning values to materials (ISO Guide 30-Ref C1). RM is necessary in method development and validation, estimation of measurement uncertainty, internal quality control, proficiency testing, and training. Both homogeneity and stability are essential in the preparation of an RM of biological origin. The homogeneity test is one of the most important steps in the development and production of a CRM [12,13]. In homogeneity studies, the various units of a material batch are important for establishing the properties of the batch. To evaluate the units, the choice of the unit should be performed randomly for that assessment to be representative of the total quantity. The National Research Council Canada (NRC) developed certified reference material (CRM) DORM-2, dogfish muscle CRM for trace metals, which was replaced by CRM DORM-3, fish protein CRM for trace metals. However, in Indonesia CRM is difficult to purchase. The objective of this study was to provide an RM which can be used as in-house reference material for quality control for the determination of total arsenic in tuna fish samples for Indonesian testing laboratories.

2. Experimental

2.1. *Preparation of a reference material*
Tuna fish samples were collected from Jakarta Bay. The tuna meat was minced, homogenized, and freeze-dried for 24 hr. Liquid nitrogen was added to the dried samples and grinded using an agate mortar. The powdered samples were sieved (100 mesh) then packed in dark glass bottles of 1 g lots.

2.2. *Tuna fish analysis*
The 50 mg of dried sample was accurately weighed. Three different methods of preparation were conducted, namely microwave assisted acid digestion, dry ashing, and waterbath acid digestion. The digest was diluted 5 times before being measured by ICP-MS. Measurements were validated using CRM DORM-2 and DORM-3.

2.3. *Homogeneity test*
Total arsenic was assessed using the different methods in two different laboratories. Twenty of 1 g lots were prepared for the homogeneity test. Random bottle number was applied in this analysis. Statistical analysis was carried out and the uncertainty of homogeneity for total arsenic was calculated using the one-way analysis of variance (ANOVA). In homogeneity testing, two importance type of methods should be noticed. Firstly, the within-bottle homogeneity, which dictates the minimum sample intake,

for which the established uncertainty is still valid. Secondly, the between-bottle homogeneity, which deals with the bottle-to-bottle or unit-to-unit variation. Separation between the heterogeneity and the variability of measurement should be established. Therefore, the within-bottle homogeneity study was carried out in a very small amount of sample; so that the between-portions effect can be quantified. When the test was completed, the minimum sample intake was calculated. On the other hand, a between-bottle homogeneity test was conducted with the optimum sample intake to minimize analytical variation. This intake should be higher than the minimum sample intake described above. When between-unit variation was able to quantify, the two effects could be separated by analysis of variance (ANOVA) [14,15].

2.4. Stability test

Two types of stability test were performed. Firstly, study at elevated temperature to elucidate whether any degradation occurred during transport. Short duration stability was carried out for not longer than 4 weeks. Secondly, study was performed at storage temperature to obtain information about the stability during storage. Usually, a stability study consists of a series of measurements performed at different times. Previous study was conducted in storage temperatures at 18 and 60°C; the reference temperature was 70°C. The storage times were 0, 1, 2, and 4 weeks. After the indicated storage periods, samples were transferred to storage at 70°C until analysis [16]. Analytical variation can be reduced by using either a method with less variation or by the isochronous measurement scheme [17], in which repeatability rather than reproducibility conditions are applied. The influence of analytical variation can be reduced further by performing higher numbers of replicates per time and by increasing the length of the stability study. In this study, the stability of total arsenic in tuna fish was evaluated over 12 months, under room temperature storage condition. Each measurement was conducted in triplicate in three different laboratories and statistical analysis was carried out using Student's t-test.

3. Results and Discussion

In the process of RM preparation, homogeneity is the first consideration. The homogeneity of a liquid or gases type of RM can be easily obtained by repetitive shaking. However, it is more difficult to homogenize a solid type of RM. Two different methods were conducted to confirm homogeneity of the candidate RM in this study.

Table 1. Results of the homogeneity test for total arsenic in the candidate reference material.

Bottle no.	ICPMS[a]	Bottle no.	ICPMS[b]
	Conc (μg/g)		Conc. (μg/g)
2	3.36	15	3.44
23	3.50	21	3.64
11	3.42	12	3.53
14	3.58	22	3.36
13	3.86	8	3.34
3	3.93	29	3.34
17	3.61	25	3.37
$\mathbf{F_{calc}}$	**3.71**		
$\mathbf{F_{crit}}$	**4.75**		

ICPMS[a] : Institut fur chemie, Karl Franzens University, Graz, Austria (microwave assisted acid digestion)
ICPMS[b] : The University of Queensland, National Research Centre for Environmental Toxicology, Brisbane, Australia (water bath acid digestion)

Table 2. Statistical calculation ANOVA single factor.

SUMMARY

Groups	*Count*	*Sum*	*Average*	*Variance*
Column 1	7	25.26	3.608571	0.046081
Column 2	7	24.02	3.431429	0.013148

ANOVA

Source of Variation	*SS*	*df*	*MS*	*F*	*P-value*	F_{crit}
Between Groups	0.109829	1	0.109829	3.708635	0.078152	4.747225
Within Groups	0.355371	12	0.029614			
Total	0.4652	13				

Another method of homogeneity testing was conducted based on ILAC Guide [14,15] using equation as follows:

$$MSB = \frac{\sum\left[(a_i + b_i) - \overline{X}_{(a_i+b_i)}\right]^2}{2(n-1)} \quad (1)$$

$$MSW = \frac{\sum\left[(a_i + b_i) - \overline{X}_{(a_i+b_i)}\right]^2}{2n} \quad (2)$$

$$Sd_{sampling} = \sqrt{(MSB - MSW)/2} \quad (3)$$

MSB = mean square between; MSW = mean square within; Sd = standard deviation

The homogeneity was measured in four criteria:

- First criteria, is the value of MSW/MSB < $F_{critical}$
- Second criteria, is the value of standard deviation sampling < 0.3 σ
- Third criteria, is the value of standard deviation sampling < 0.3 x CV Horwitz prediction
- Fourth criteria, is the value of standard deviation sampling < CV Horwitz prediction

From the data above, the MSB and MSW value were found to be 0.0140 and 0.0388, respectively. The F statistic value was calculated from MSW/MSB = 2.77, this value was lower than F critical from F table (at α= 0.05) of 3.71. In this study, the RM was homogeny.

Table 3. Results of stability test of total arsenic in the candidate reference material.

	Total arsenic concentration (μg/g)						
	Dec	Feb	April	June	August	Oct	Dec
	3.53	3.18	3.77	3.47	3.44	3.44	3.44
	3.31	3.54	3.69	3.78	3.12	3.41	3.64
	3.44	3.12	3.89	3.71	3.46	3.39	3.53
Average	3.43	3.28	3.78	3.65	3.34	3.41	3.54
Standard deviation	0.11	0.23	0.10	0.16	0.19	0.03	0.10
t_{calc}		**2.13**	**1.32**	**0.32**	**1.87**	**1.52**	**0.54**

Beside homogeneity, the stability of the RM is important. In chemical testing laboratories, dried samples are usually stored at room temperature. Therefore, the stability test for this candidate RM was performed over 12 months at room temperature and measured in three different laboratories. The statistical results showed no significant changes in stability as measured by the total arsenic concentration, because t_{calc} were lower that t_{crit} for the degree of freedom of n-2 and confidence level of 95%, as shown in Table 3.

The concentration of total arsenic was found steady for 12 months as shown in Figure 1, with the linear equation of y = 0.0027x + 3.4739.

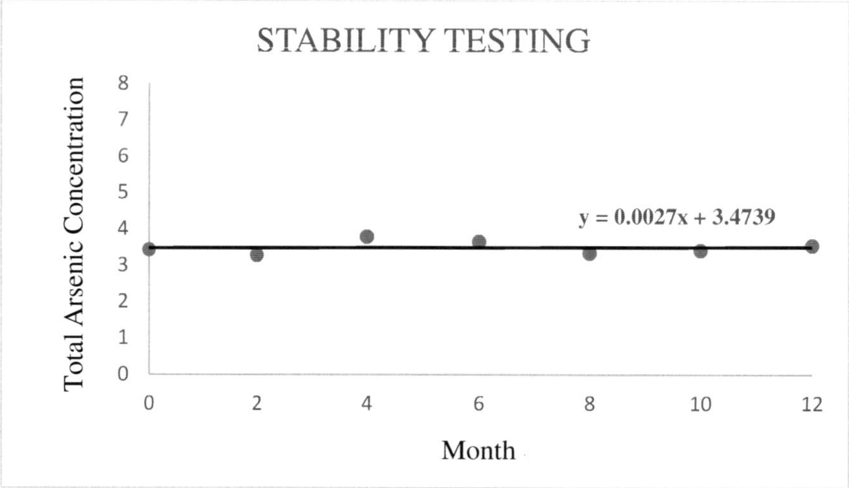

Figure 1. Total arsenic concentration in 12 months storage.

CRM DORM-2 and DORM-3 were periodically analyzed for more than nine years. It was found physically and chemically stable over that time period [18,19].

4. Conclusions

No significant difference was observed among aliquots of the candidate RM. It also showed that there was no significant changes in stability of the candidate material over 12 months at room temperature. The candidate RM developed in the present study is suitable for quality assurance for the determination of total arsenic in tuna fish as well as satisfactory for its homogeneity and stability. The candidate RM also demonstrated its suitability for quality assurance of the total arsenic measurement in tuna fish.

Acknowledgments

We would like to thank Sandwich Program Scholarship of The Ministry of Science and Technology and The Directorate of Higher Education of Indonesia for its financial funding.

References

[1] Hughes M F, Beck B D, Chen Y, Lewis A S and Thomas D J 2011 *Toxicol. Sci.* **123** 305–32
[2] Rodríguez-Lado L, Sun G, Berg M, Zhang Q, Xue H and Zheng Q 2013 *Science* **341** 866–8
[3] ISO/IEC 17025:2005 *General Requirements for the Competence of Testing and Calibration Laboratories* (Geneva: International Organization for Standardization)
[4] ISO GUIDE 34:2000 (E) *General Requirements for the Competence of Reference Material Producers 2nd ed.* (Geneva: International Organization for Standardization)
[5] ISO GUIDE 35:2006 (E) *Reference Materials – General and Statistical Principles for Certification 3rd ed.* (Geneva: International Organization for Standardization)
[6] Koesmawati T A, Buchari B, Amran M B and Kardono L B S 2013 *J. App. Phar. Sci.* **3** 116-21

IOP Conf. Series: Earth and Environmental Science **160** (2018) 012022 doi:10.1088/1755-1315/160/1/012022

[7] Penrose W R, Black R and Hayward J 1975. *J. Fish. Res. Board Can.* **32** 1275-81
[8] Whaley-Martin K J, Kock I and Reiner K J 2012 *Talanta* **88** 187-92
[9] Leufroy A, Noel L, Dufailly V, Beauchemin D and Guerin T 2011 *Talanta* **83** 770-9
[10] IUPAC (International Union of Pure and Applied Chemistry) 2002 **74** 835–55
[11] AOAC 1999 *Guidelines for Single Laboratory Validation of Analytical Methods for Trace-level Concentrations of Organic Chemicals* (Washington: AOAC)
[12] dos Santos A M P, dos Santos L O, Brandao G C, Leao D J, Bernedo A V B, Lopes R T and Lemos V A 2015 *Food Chem.* **178** 287–91
[13] Shen Y and Chen C 2007 *Biosyst. Eng.* **97** 41-9
[14] ILAC Proficiency Testing Consultative Group 2000 *ILAC Discussion Paper on Homogeneity and Stability Testing*
[15] Horwitz W and Albert R J 2006 *Assoc. Off. Anal. Chem.* **89** 1095-109
[16] Zeleny R 2015 *Food Chem.* **168** 241–6
[17] Linsinger T P J, Pauwels J, van der Veen A M H, Schimmel H and Lamberty A 2001 *Accred. Qual. Assur.* **6** 20–5
[18] DORM-2-Dogfish Muscle Certified Reference Material for Trace Metals 1993 *National Research Council Canada (NRCC)*
[19] DORM-3-Fish Protein Muscle Certified Reference Material for Trace Metals 2007 *National Research Council Canada (NRCC)*

IOP Conf. Series: Earth and Environmental Science **160** (2018) 012023 doi:10.1088/1755-1315/160/1/012023

Residual pesticides monitoring of the horticulture soil in Lembang

R Yusiasih*, **Efendi, M M Pitoi and H R Hariyadi**

Research Unit for Clean Technology, Indonesian Institute of Sciences (LIPI)
Komplek LIPI Bandung, Jalan Sangkuriang, Gedung 50, Bandung 40135, Indonesia

*Email: retno.yusiasih@lipi.go.id

Abstract. The use of pesticides in agricultural systems is one of the most important factors which contributes to the massive increase of food production worldwide. In this study, spike technical approach was carried out to simulate the concentration of pesticide. The sample was taken from soil in Lembang region. The pesticide from the spiked soil was extracted with sonication technique and analyzed based on HPLC method. The study showed that 10 gram positive and 2 gram negative bacteria isolates were found from the Lembang soil. The inhibition concentration (IC_{50}) of pesticides on microbial growth was 50%.

1. Introduction

Pesticides have been used extensively since the compounds are beneficial to agriculture as well as its usage greatly increases from year to year [1]. The majority of such substances are applied directly to soil or sprayed over crop fields and hence released directly to the environment. Therefore, pesticides contaminate soil surface of farmland [2]. The slow degradation in the environment leads to environmental contamination and adverse effects to humans. Pesticides in soils are studied more than any other environmental contaminants. The monitoring of their residues in soils were reported [3].

Sample preparation prior to the determination of organic pesticides in soil usually consists of several steps due to the complexity of the matrix. The sample preparation step can be marked as the most critical. The isolation of pesticides from soil samples usually involves conventional solid–liquid extraction methods such as soxhlet extraction. However, these methods are time consuming, labor intensive, and demand large volumes of hazardous organic solvents with high cost of both purchase and disposal. The modern sample preparation procedures, such as supercritical fluid extraction (SFE), matrix solid-phase dispersion (MSPD), headspace solid-phase micro extraction (HS-SPME), ultrasonic-assisted extraction (UAE), and microwave-assisted extraction (MAE) have been developed to overcome the drawbacks of the traditional approaches. Various studies reported that ultrasonic extraction of organic analytes from soil and sediment are effective methods. However, the application of ultrasonication to extract the carbofuran from soil is rarely studied [2-6].

Pesticides help controlling pests for a good cropping and yield, but they also create several holistic anomalies in soil health as residual effects [7]. The soil contamination of pesticides may impact the soil microorganism by inhibit their growth as well as their activities [8,9]. Bacterial diversity in soil is altered by pesticide use [10] along with its population number [11]. In addition, decreasing in enzyme activity is also observed under pesticides use [12,13].

This research was aimed to monitor the concentration of carbofuran as pesticide residue in soil taken from agricultural area in Lembang, West Java. Bacteria isolated from Lembang soil based on type of gram staining was characterized. The inhibition effect of carbofuran on the growth of indigenous microbes was also observed.

2. Experimental

2.1. Chemicals and reagents
Carbofuran was obtained from Hunan Research Institute of Chemical Industry-PR. China with 99.7% of purity. This pesticide was used for 100 mg/L stock standard solutions in methanol and stored at 4°C in the dark. Methanol and acetonitrile were HPLC grade from Merck. Deionized water was obtained from PT. Ikapharmindo Putramas. Membrane filter of 0.45 μm was used for filtration. Nutrient agar from Difco was used as growth medium for bacterial isolation. CH_3COONa, KH_2PO_4, $Na_2HPO_4.7H_2O$, and NH_4Cl used for toxicity test were purchased from Merck.

2.2. Apparatus
An ultrasonic water bath Branson-3510, centrifuge EBA-12 and vortex Genius-3 were used for extraction. An HPLC Hitachi-7000 series was used for the carbofuran analysis, which was performed on a μ-Bondapak C_{18} column with UV detector set at 275 nm. The mobile phase was acetonitrile and deionized water with the ratio of 1:1 (v/v). The flow rate of eluent was 0.8 mL/min with an isocratic mode. The analysis was performed at ambient temperature and the injection volume was 20 μL. The spectrophotometer Agilent Technologies Carry 60 UV/VIS was used for the optical density as microbial growth indicator was measured at 540 nm.

2.3. Source, collection and soil sample treatment
The soil samples were collected from Lembang farming area, West Java, Indonesia, from 0 - 20 cm under the vegetable plantations, then stored at 4°C. Soil samples were homogenized, sieved to obtain the same particle size and air-dried at room temperature before use. The control soil of which free from carbofuran was spiked with carbofuran to achieve its concentration to 5 mg/kg. The soils were then kept overnight at room temperature.

2.4. Extraction of carbofuran in soil sample
Extraction of carbofuran in soil was carried out by sonication technique. The 5 g of the spiked soil was extracted with 20 mL of acetonitrile for 1, 2, and 3 h of sonication. The soil samples were then centrifuged at 4000 rpm for 5 min. The extract then was filtered through 0.45 μm syringe filters and the filtered solutions were analyzed by HPLC. The blank samples were prepared with the same procedure but without carbofuran spiking.

2.5. Isolation of bacteria in soil
To isolate the bacteria, several suspensions were made. Suspension 1 was made by mixing 0.5 g of soil sample and 10 mL of 0.9% salt solution mixed with 0.05% tween 80. Suspension 2 was made by mixing homogeneously 1 mL of suspension 1 and 9 mL of salt solution. Suspension 3 was made similarly, by mixing homogeneously 1 mL of suspension 2 and 9 mL of salt solution. The suspension 3 was then used to culture the bacteria in which 1 mL of the suspension was added into each sterile petri dish followed by pouring a nutrient agar media at 40-50°C. The petri dish was then closed tightly and incubated for 1-2 days at 37°C. Bacteria purification was done by tilting the petri dish and incubated for 6 days. Bacteria isolation was done through pour plate method.

2.6. Inhibition pesticides on microbial growth (IC_{50})
Inhibition concentration of pesticides on microbial growth by 50% was observed to determine the toxicity of the compound based on microbial assay. The toxicity of pesticides was carried out based on the method develop by Alsop *et al* [14]. The medium (150 mL) contained CH_3COONa, 1500 mg/L; KH_2PO_4, 850 mg/L; K_2HPO_4, 950 mg/L; Na_2HPO_4 $7H_2O$, 1500 mg/L, NH_4Cl; 850 mg/L and pesticides

(0-1000 mg/L). After inoculated with 1 mL microbial inoculum enriched from noncontaminated soil, the cultures were incubated at 30°C with orbital shaking (100 rpm) for 20 h. The optical density as growth indicator was then measured at 540 nm using a spectrophotometer.

3. Result and discussion

3.1. Qualitative analysis on spike soil samples

Extraction of carbofuran in soil reveals good results as shown in Figure 1. In the chromatogram, peak of carbofuran from the standard solution was appeared at the retention time (t_R) of 8.07 min (A), whereas carbofuran from the spiked soil sample at 8.04 min (C). The relative percent difference (RPD) both of the t_R was 0.37%, and the blank soil sample showed no peak of carbofuran (B).

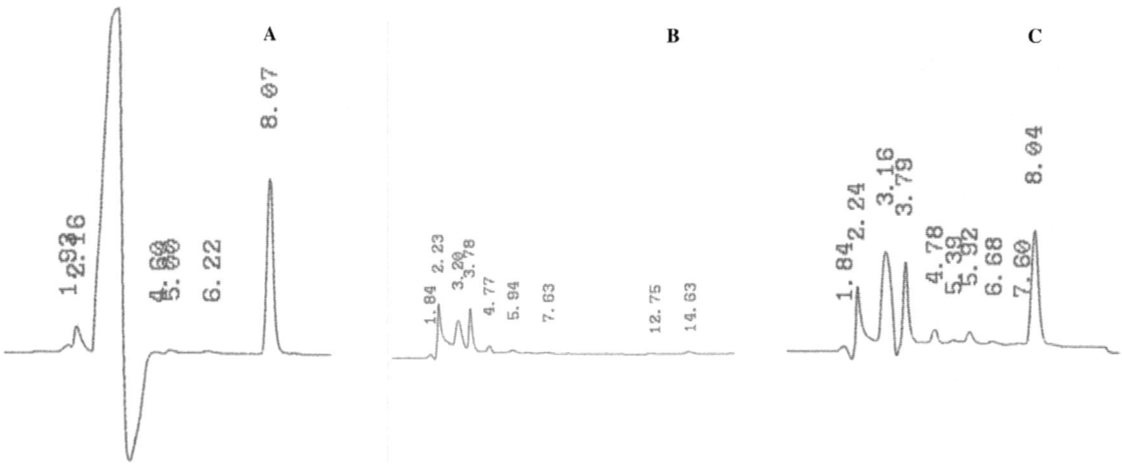

Figure 1. Chromatograms of (A) carbofuran standard solution, (B) blank soil sample, and (C) spiked soil sample.

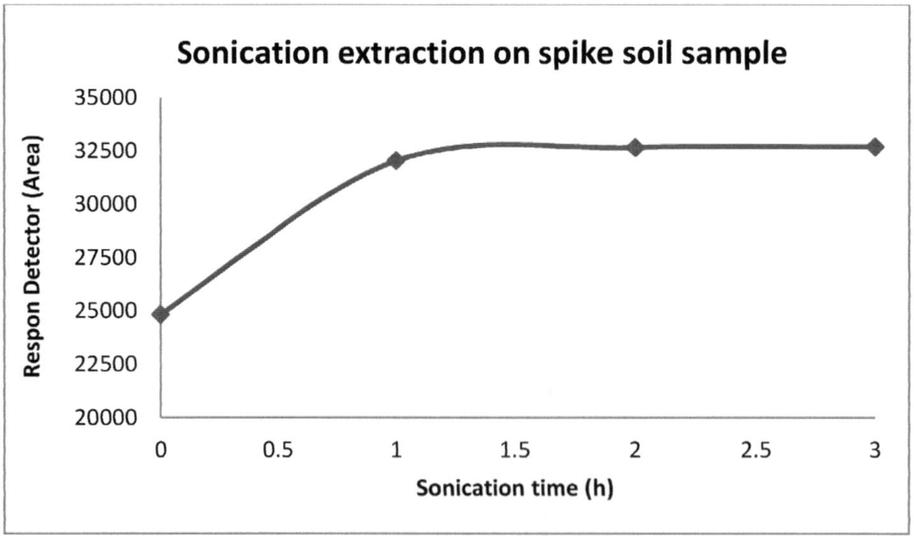

Figure 2. Sonication effect on spike soil sample extraction.

Parameter of the sonication efficiency such as time of the sonication was also investigated and the result can be seen in Figure 2. At the 5 mg/kg fortification levels of carbofuran in 5 g soil sample, the extraction was optimum at 1 h sonication with only 20 mL of acetonitrile as a solvent. The analytes in the extracts were analyzed directly by HPLC-UV without any further cleanup. The recovery of extraction was in the range of 109 to111%. Extraction by sonication technique was simple, time- and cost-effective. The concept of green analytical chemistry (GAC) has been done. The key goals to be achieved in greening analytical methods are elimination or reduction of reagents, waste, risk and hazard [15].

3.2. Bacterial isolation in soil samples

The Lembang horticultural soil analysis was carried out on moisture content and bacterial types. The moisture content was conducted by gravimetric method and its value was 31.5%. The bacterial isolate was shown in Figure 3. The composition of bacteria communities were 2 isolates of gram negative and 10 isolates of gram positive. The five carbamate insecticides *in vitro* and their average acute toxicity levels to the cyanobacteria were in a descending order: carbaryl > carbofuran, propoxur and metolcarb > carbosulfan [7].

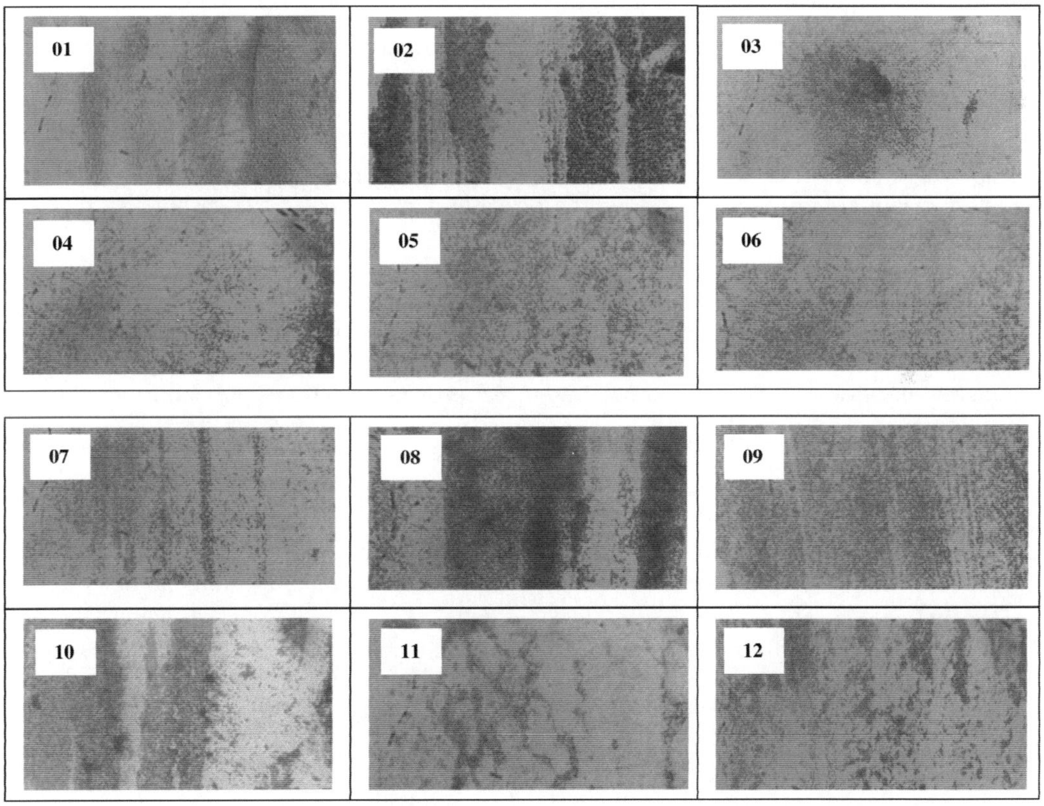

Figure 3. Bacterial types of Lembang soil: (01-02) gram negative and (03-12) gram positive.

3.3. Inhibition pesticides on microbial growth (IC_{50})

The results of the toxicity test are shown in Figure 4. The growth of microbes compared to the control was not significantly inhibited at pesticide concentration of 0.1 to 10 mg/L. This was shown by a high value of optical density of which was 87% at 10 mg/L of carbofuran. However, at 100 mg/L, the inhibition of the microbes observed with the optical density was 33% compared to the control.

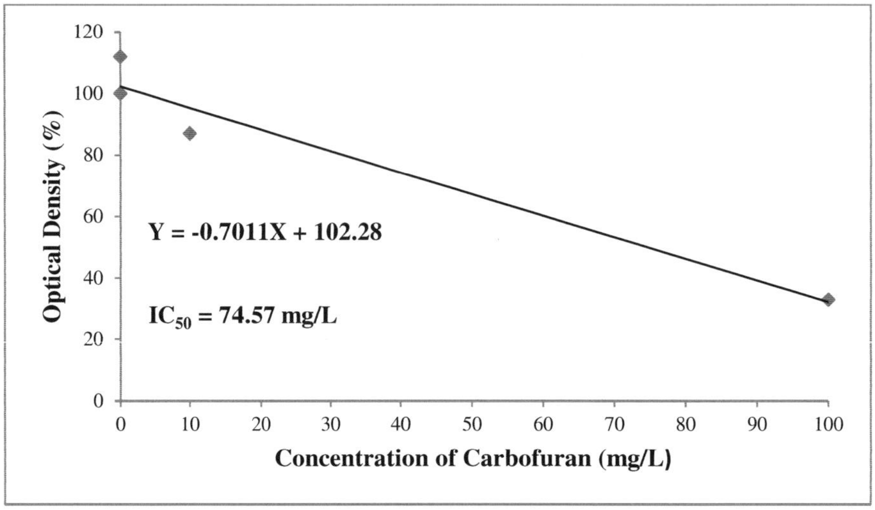

Figure 4. Toxicity test of carbofuran based on microbial growth.

The inhibition of microbial growth by 50% compared to the control was observed at the pesticide concentration between 10 and 100 mg/L. The IC_{50}, which is the concentration that inhibited the microbial growth by 50% of the control culture with no added pesticide, was 74.57 mg/L. Carbofuran is highly toxic after a single oral dose; the LD_{50} values in rats range from 6 to 18 mg/kg bw whereas and in other species such as mouse, guinea-pig, rabbit, cat and dog, ranges from 3 to 19 mg/kg bw [16]. Further study was recommended to observe the possible degradation of the carbofuran by microbial association.

4. Conclusion

From the current study, it is concluded that with the moisture content of was 31.5%, horticulture soil in Lembang contained 2 isolates of gram negative bacteria and 10 isolates of gram positive bacteria. The inhibition concentration of carbofuran on the microbial growth by 50% of the control culture without pesticide addition was 74.57 mg/L.

Acknowledgements

This work was supported financially by Indonesian Institute of Sciences under the scheme of Penelitian Mandiri 2017. The authors wish to thank Miss Ekbar Pauti Sutia, Miss Hanadia Hielwana and Miss Nurjihan D. Qur'ani for their assisstance to carry out the bacterial isolation and the toxicity test.

References

[1] De A, Bose R, Kumar A and Mozumdar S 2014 *Targeted Delivery of Pesticides Using Biodegradable Polymeric Nanoparticles* (New Delhi: Springer) pp 5-6
[2] Vega A B, Frenich A G and Vidal J L M 2005 *Anal. Chem. Acta* **538** 117–27
[3] Merdassa Y, Liu J and Megersa N 2013 *Talanta* **114** 227–34
[4] Tor A, Aydin M E and Özcan S 2006 *Anal. Chim. Acta* **559** 173–80
[5] Đurović R and Đorđevic T 2011 *Pesticides in the Modern World - Trends in Pesticides Analysis* ed M Stoytcheva (Belgrade: Institute of Pesticides and Environmental Protection) pp 221-46
[6] Arias J L O, Rombaldi C, Caldas S S and Primel E G 2014 *J. Chromat. A* **1360** 66–75
[7] Padhy R N and Rath S 2015 *Rice Sci.* **22** 89-98
[8] Imfeld G and Vuilleumier S 2012 *Eur. J. Soil Biol.* **49** 22-30
[9] Omar S and Abdel-Sater M 2001 *Water Air Soil Pollut.* **127** 49-63
[10] Lo C-C 2010 *J. Environ. Sci. Health B* **45** 348-59
[11] Johnsen K, Jacobsen C S, Torsvik V and Sørensen J 2001 *Biol. Fert. Soils* **33** 443-53

[12] Floch C, Chevremont A-C, Joanico K, Capowiez Y and Criquet S 2011 *Eur. J. Soil Biol.* **47** 256-63
[13] Jacobsen C S and Hjelmsø M H 2014 *Curr. Opin. Biotechnol.* **27** 15-20
[14] Alsop G M, Wage G T and Coney R A 1980 *J. Water Pollut. Con. F.* **52** 2452-6
[15] Yusiasih R, Nugraha W C and Hudayya A 2017 *IOP Conf. Ser.: Earth Environ. Sci.* **60** 012024
[16] Food and Agricultural Organization 2009 *Pesticide residues in food 2008-Joint FAO/WHO meeting on pesticide residues* (Rome, Italy: Food and Agricultural Organization of the United Nations) pp 123-6

IOP Conf. Series: Earth and Environmental Science **160** (2018) 012024 doi:10.1088/1755-1315/160/1/012024

Preparation of copper amalgam (CuHg) as working electrode for analysis of selenium

F Dara[1*], Buchari[2] and I Noviandri[2]

[1] Research Unit for Clean Technology, Indonesian Institute of Sciences (LIPI) Komplek LIPI Bandung, Jalan Sangkuriang, Gedung 50, Bandung, 40135, Indonesia
[2] Analytical Chemistry Research Group, Chemistry Department, Institut Teknologi Bandung Jalan Ganeca No.10, Bandung 40132, Indonesia

*Email : fitr009@lipi.go.id; fitridara74@gmail.com

Abstract. Selenium is an essential trace element for human. Beside the range of concentration, the beneficial and toxic effects of selenium are also depending on which chemical form it takes. In nature or environment and some organisms, selenium is present both in organic and inorganic forms where some studies revealed that inorganic species especially selenite are more toxic than the organics matter. Various methods are used to analyse selenium, but it is usually to determine the total selenium, not designed for field applications, requires several preparation steps prior to actual determination, and needs a lot of chemicals. These problems can be solved by voltammetry methods by using dropping or film mercury as a working electrode. The nature of poisonous mercury becomes a limitation in the use of this working electrode. A copper amalgam (CuHg) as working electrodes was successfully prepared for the analysis of selenium. CuHg working electrode was made in a simple way, namely by deposition of copper (Cu) on the surface of Cu wire with potential and at certain time, then dipped into a solution of saturated mercury salt to form the amalgam. Cyclic voltammetry was used to investigate the electrochemical properties of selenium with 0.1 M HCl as a supporting electrolyte. Cyclic voltammetry technique was conducted on the working potential range of -300 mV (E_i) to -800 mV (E_f). Profiles of cyclic voltammogram were influenced by the scan rate applied. In addition, SEM images showed that the adsorption of selenium occurred on the surface of CuHg working electrode.

1. Introduction

Selenium (Se) is an essential trace element for biological system, but it will be as a toxic if it presents excessively. The toxic effects of selenium is strongly depending on its chemical form. In general, inorganic selenium compounds are more toxic than the organic ones [1]. The most common forms of selenium in water are selenite (Se(IV), SeO_3^{2-}) and selenate (Se(VI), SeO_4^{2-}). The formation of selenate from selenite is slow, and both forms exist together in solution [2]. Selenite is approximately 10-fold more toxic than selenate [3].

Atomic absorption spectrometry with hydride generation is one of the powerful methods for selenium determination due to its low detection limit. The method is inconvenient and normally not available in several laboratories in Indonesia because it is expensive, not easy for actual measurements, require previous preparation steps, and needs a lot of chemicals.

Electrometric is widely diffused as alternative methods due to the capability for concentration evaluation quantitatively up to the trace levels as well as the instrumentation is relatively cheap. Cyclic voltammetry, one of electrometric methods, is carried out for the initial electro analytical studies for any compound. This method is carried out to characterize the electrochemical processes of the compounds before investigated in depth using other techniques [4]. Analysis of selenium using copper amalgam as working electrode has been investigating previously [5-8]. This paper reports the cyclic voltammetry study of selenite ion using copper amalgam (CuHg) as working electrode.

2. Experiment

2.1. Apparatus

All measurements were performed using an eDAQ potentiostat model EA163 with an e-corder unit connected to a PC, in combination with a three-electrode cell incorporating platinum wire (Nilaco, Japan) as the auxiliary electrode, Ag/AgCl electrode (3 M KCl) as reference, and copper amalgam (CuHg) as working electrode.

2.2.Chemicals and reagents

All chemicals used were of analytical grade. Cu wire was obtained from PT. Freeport Indonesia. Copper sulfate pentahydrate, hydrochloric acid 37%, and mercury nitrate were purchased from E.Merck. Sodium selenite pentahydrate 99% was purchased from Sigma Aldrich, Japan. Ultra-pure water (resistivity 18.2 $M\Omega cm^{-1}$) was prepared using Milli-Q water systems, Millipore-waters USA at Research Centre for Chemistry, Indonesian Institutes of Sciences.

3. Procedures

3.1. Copper amalgam (CuHg) preparation

Copper amalgam CuHg as working electrode was made by deposition of Cu from $CuSO_4$ 0.1 M at potential deposition 15 V for 35 s onto Cu wire with diameter 0.4 cm, then immersed in saturated mercury salt. Copper amalgam (CuHg) electrode was sequentially cleaned with ultra-pure water and acetone. It was then dried at room temperature.

3.2. Standard solution preparation

Sodium selenite pentahydrate of ± 0.05 g was diluted with 100 mL of 0.1 M HCl in volumetric flask for selenite standard solution. More dilute standard solution was prepared by weighing and dilution with 0.1 M HCl. These standard solutions were prepared prior to its use.

3.3. Cyclic voltammetry of selenite ion

Cyclic Voltammetry (CV) of selenite ion was studied using 0.1 M HCl media as supporting electrolyte refer to previous literatures [9-11]. The copper amalgam (CuHg) and Ag/AgCl were used as the working and reference electrode, respectively. The general procedure used to obtain cyclic voltammograms was as follows: An amount of 5 mL selenite ion solution 1 mM was placed in a voltammographic cell and scanned from -300 to -800 mV with potential accumulation -450 mV and accumulation time 600 s.

4. Result and discussion

At the copper amalgam (CuHg) electrode, electrodeposition of selenite ion 1 mM was characterized by cathodic and anodic peak potential at *ca* -700 mV and -600 mV (vs Ag/AgCl), respectively. The cyclic voltammogram is shown in Figure 1A. When the concentration of selenite ion was 50-fold lower than 1 mM, anodic peak could not be found (Figure 1B) which indicates that the selenite ion reduction was not totally reversible but quasi-reversible.

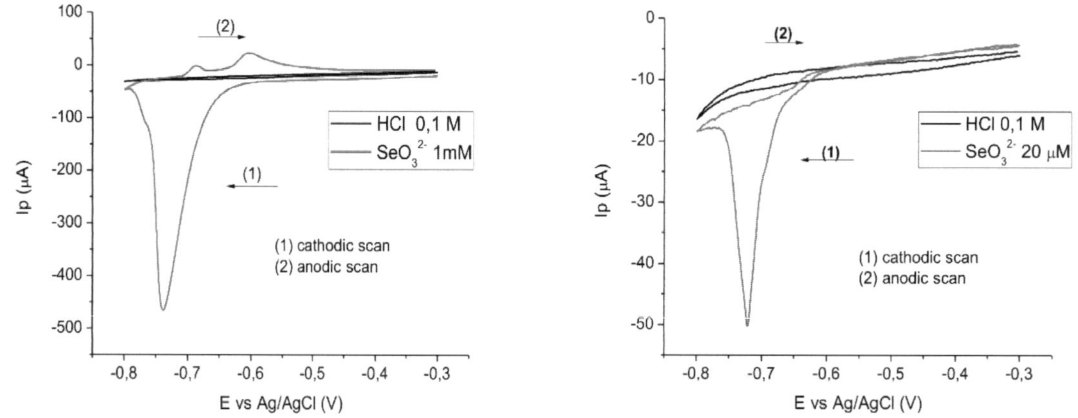

(A) (B)

Figure 1. Cyclic voltammogram of selenite ion at optimum accumulation time and potential. (A) 1 mM in 0.1 M HCl (B) 20 µM in 0.1 M HCl.

To confirm that the observed peaks were related to the reduction of selenite ion, CV was carried out with different selenite ion concentration. Figure 2 shows that cathodic peak at *ca* –700 mV (vs Ag/AgCl) was referred to the reduction peak of selenite ion since Ipc increased with increasing of selenite ion concentration.

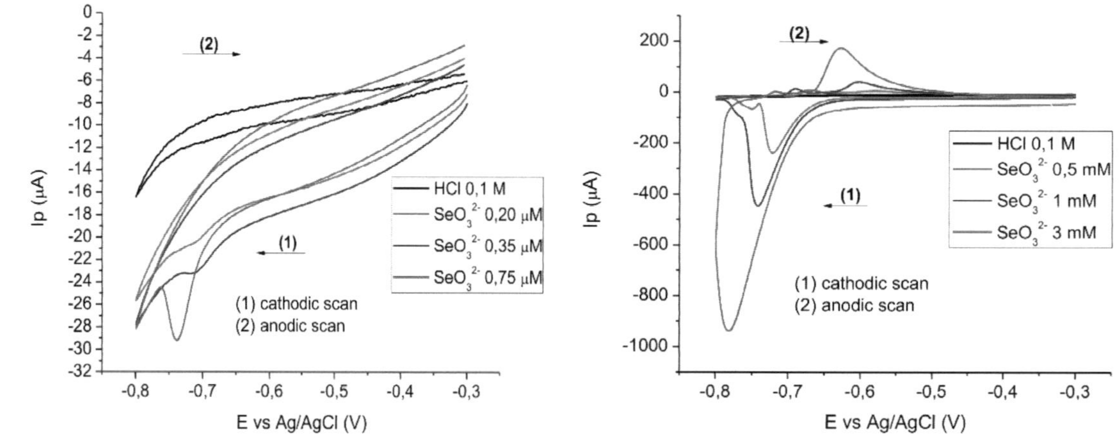

Figure 2. Cyclic voltammogram of different concentration of selenite ion in 0.1 M HCl.

The effect of increasing scan rate (from 10 to 200 mV/s) to potential and current peak of reduction selenite ion were observed under the same experiment conditions. As the scan rate of CV was increased, the reduction peak Ipc increased and shifted more negative. Figure 3 shows the CV of selenite ion in 0.1 M HCl recorded at different scan rate. This result indicates that there was a linear relationship between the scan rate (v) and the peak current with correlation coefficient r^2= 0.95745 as shown in Figure 4A. The linear regression equation is given by: Ipc (µA) = -0.4667v - 3.8205 and Ipa (µA) = 0.1076v + 2.2495. On the other hand, a linear relationship between the square root of the scan rate (v½) and the peak current with correlation coefficient r^2= 0,97039 as shown in Figure 4B was also observed. The linear regression equation is given by: Ipc (µA) = -9.1337v + 36.7444 and Ipa (µA) = 2.3001v - 9.4288. This implies that the mechanism was controlled by a mixture of both adsorption and diffusion.

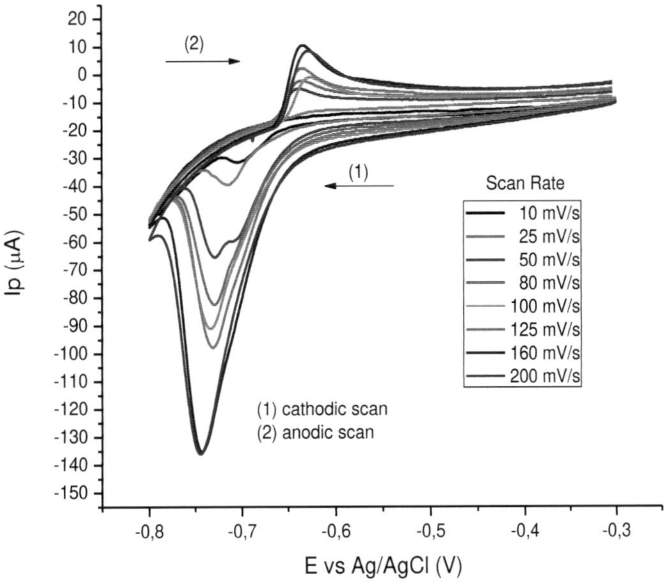

Figure 3. Cyclic voltammogram of selenite ion 20 µM in 0.1 M HCl recorded at different scan rate.

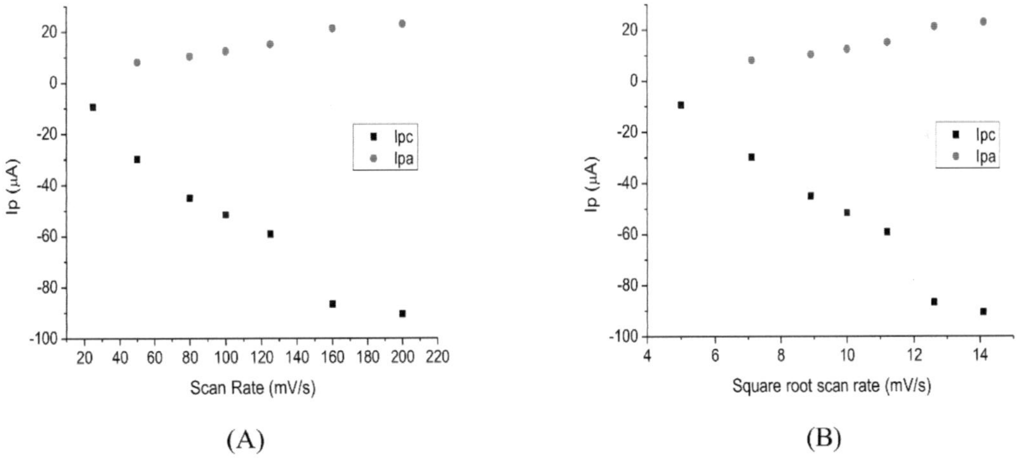

Figure 4. (A) Plot of the cathodic and anodic peak current of selenite ion 20µM in 0.1 M HCl as a function of scan rate.
(B) Plot of the cathodic and anodic peak current of selenite ion 20µM in 0.1 M HCl as a function of the square root of the scan rate.

5. Conclusion

The characteristic electrochemical processes of the selenite ion in 0.1 M HCl on the copper amalgam (CuHg) electrode have been investigated. The reduction process is a quasi-reversible reaction with a characteristic cathodic and anodic peak at *ca* –700 mV and *ca* -600 mV (vs Ag/AgCl), respectively. The limiting current is controlled by diffusion and adsorption process.

Acknowledgement
The authors would like to thank Kementrian Riset dan Teknologi – Dikti Indonesia for financial support.

References

[1] Ferri T, Petruzzelli G, Pezzarossa B, Santaroni P, Brunori C and Morabito R 2003 *Microchem. J.* **74** 257-65
[2] World Health Organization 2011 *Guidelines for Drinking Water Quality 4th ed.* (Geneva: World Health Organization)
[3] Lemy A D, Finger S E and Nelson M K 1993 *Environ. Toxicol. Chem.* **12** 2265-79
[4] Wang J 2000 *Analytical Electrochemistry 2nd ed.* (New York: Wiley-VCH)
[5] Yosypchuk B and Novotný L 2002 *Crit. Rev. Anal. Chem.* **32** 141-51
[6] Yosypchuk B and Barek J 2009 *Crit. Rev. Anal. Chem.* **39** 189-203
[7] Piech R 2008 *Electroanalysis* **20** 2475-81
[8] Ashournia M and Aliakbar A 2009 *J. Hazard. Mat.* **168** 542-7
[9] Bryce D W, Izquierdo A and Castro M D 1995 *Fresenius J. Anal. Chem.* **351** 433-7
[10] Piech R and Kubiak W W 2007 *Electrochim. Acta* **53** 584-9
[11] Tanaka S, Sugarawa K and Taga M 1990 *Anal. Sci.* **6** 475-8